EVOLUTION:
When Fact Became Fiction

EVOLUTION:
When Fact Became Fiction

by Ricki D. Pavlu, M.S.

Evolution: When Fact Became Fiction

by Ricki D. Pavlu, M.S.

Hazelwood, Mo. 63042

Illustrated by Anthony Wimberly

Cover Design by Tim Agnew

All Scripture quotations in this book are from the King James Version of the Bible unless otherwise identified.

Printed in United States of America.

Printed by

WAP

Library of Congress Cataloging-in-Publications Data

Pavlu, Ricki D., 1952-
Evolution : when fact became fiction.

Bibliography: p.
1. Bible and evolution, 2. Evolution—Controversial literature.
3. Creation. I. Title.
BS659.P38 1986 231.7'65 86-13144
ISBN 0-932581-51-X

Table of Contents

Foreword

I have been asked to open the front door of this book. I will not keep you long on the threshold, for between its pages are some exciting facts. Rivers of ink and mountains of paper have been used to write articles concerning the theory of evolution versus the truth of creation. In my opinion, none has excelled in layman's language any better than this effort by Reverend Ricki Pavlu.

You will find his insight factual, keen, informative, and most of all documented. Reverend Pavlu has done the Christian world a great service by pouring himself into this work.

It is especially palatable to the ordinary layman living on the periphery of scientific knowledge. As you peruse its pages, you will find yourself quite capable of coming to grips with all of its profound presentation. This hallmark grants the book a distinctive place among other efforts in this area. Frankly—let me put it another way—you can understand what the man is saying!

Get into it, read it through, and then keep it as a reference. I especially commend it to young people who are swimming in a sea of confusion today. Every generation must get its personal bearings. They must hear what the Scriptures speak to them in their given time. This book is a great asset. Now that you are past the threshold, get right into the living room and work your way to the back page. You will find the Light is shining there, too.

Tom Fred Tenney

Preface

The professor stood staring at me, his mouth agape. How could a twenty-year-old sophomore college student not only contradict him, a college professor, but also challenge him with questions for which he had no answers? After all, had not he so often asked problematic questions of Christian students and had watched them slink back in their chairs, their faces red with embarrassment? He now realized that I had been doing some research of my own concerning the so-called "fact" of evolution. That was in the fall of 1972 at Louisiana Tech University in Ruston, Louisiana. Since that time, my research continued and is now compiled in this book.

The central goal of this book is to make the reader vividly aware that he or she does not have to accept the theory of evolution as fact. The alleged evidence for evolution is, in reality, no evidence at all. Next, it is hoped that the material presented will solidify the Christian's confidence in God as the Creator of the universe.

This book is written in such a manner that the average reader can follow the reasoning and argumentation. As with any technical book, scientific terms and concepts have been used. Extreme care has been taken, however, to define these terms and concepts in such a manner that readers from all walks of life can understand what is being said.

I gratefully acknowledge the encouragement of Reverend David Hale, who first gave me the idea of writing this book. I would also like to thank Reverend John Burns, Charland Jordan, Gary Moore, and James M. Wright for reviewing the entire manuscript. Words could not express my thanks to Ms. Audrey C. Harbin for editing the manuscript a countless number of times and for typing and retyping the manuscript. Special thanks goes to Reverend T. F. Tenney and his wife

Thetus for writing the Foreword and contributing helpful suggestions. I am also indebted to Mr. Anthony Wimberly for the artwork and illustrations.

Chapter 1
THE WORLD THAT THEN WAS

"Whereby the world that then was, being overflowed with water, perished" (II Peter 3:6).

If we could step a few thousand years into the past and view the world that Noah lived in before the Flood, it would certainly be a different world from ours today. To understand Noah's world fully, however, we must regress even one step farther back to the world that Adam and Eve lived in before the curse.

The World Before the Curse

According to Genesis 1:31, everything God made was perfect. God created perfect human beings (male and female) and placed them in a perfect paradise where everything existed in perfect harmony. There was no pain or suffering. According to Genesis 1:29-30, both man and beast were vegetarians, so neither killed the other for food. Death had not entered the world because Adam had not sinned (Romans 5:12). Rain and its erosive leaching effects were non-existent; rather, a mist pushed it's way through the earth carrying with it nutrients essential to the production of faultless herbs and fruit-bearing trees (Genesis 2:5-6).

Words could never convey the beauty of this paradise and the perfect harmony man had with his environment and with his God. Had it not been for sin, man would be a part of that astounding world today.

The Curse

"And unto Adam he said, Because thou hast hearkened unto the voice of thy wife, and hast eaten of the tree, of which I commanded thee, saying, Thou shalt not eat of it: cursed is the ground for thy sake; in sorrow shalt thou eat of it all the days of thy life; Thorns also and thistles shall it bring forth to thee; and thou shalt eat the herb of the field; in the sweat of thy face shalt thou eat bread, till thou return unto the ground; for out of it wast thou taken: for dust thou art, and unto dust shalt thou return" (Genesis 3:17-19).

We should notice that God placed a curse upon the ground because of man's sin. This brings us to the essential conflict between the two philosophies of creation and evolution. Evolution advocates that man is improving and will soon approach Utopia, whereas the Bible says man has fallen and is still degenerating.

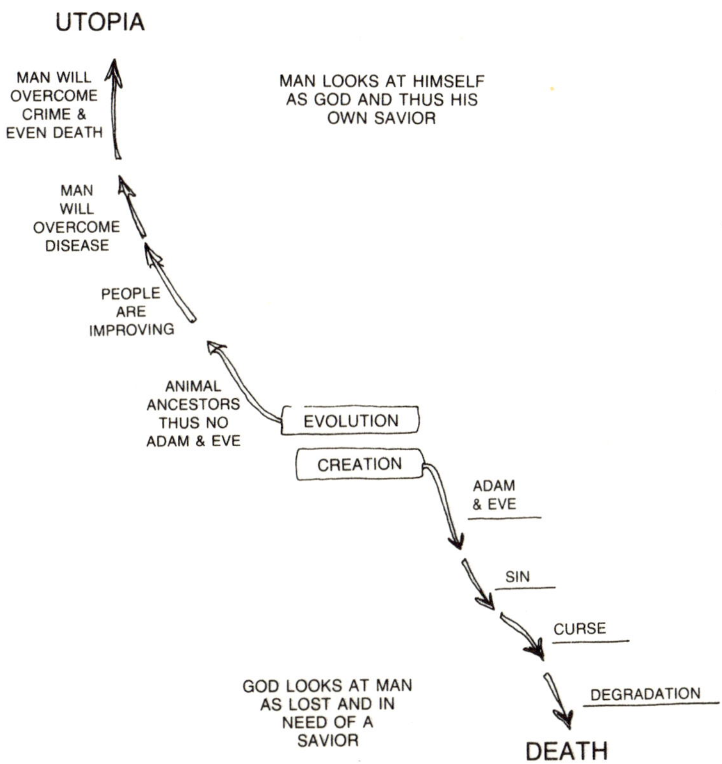

Many Christians today are unconcerned about the doctrine of evolution. In their view, if scientists can eventually prove evolution to be true, then it would only mean this was the method by which God created the universe. In so doing, as the preceding illustration plainly shows, these people are inadvertently denying the inspiration of the Bible. The question is, "Shall we listen to the opinion of fallen man or to the inspired Word of a perfect God?"

The Bible teaches that there was an initial creation which set the universe in faultless order. This included the creation of a perfect earth with perfect fauna (animals) and flora (plants). This model world continued until the Law of Universal Decay (the curse God placed upon the entire creation, which scientists call the Second Law of Thermodynamics) took effect. From this point in history onward, the universe and everything in it has been gradually decaying. This decaying process will continue until God removes the curse (Revelation 22:3).

The Second Law of Thermodynamics

In simple terms, the Second Law of Thermodynamics states that all systems have a tendency toward increased disorder, which is called "entropy." In other words, everything in the universe is slowing down and wearing out. Every star, including our own sun, is gradually burning out. Everything in nature and all things that men have put together are falling apart and wearing out. Man builds a house to see it rot away. He assembles a car to see it rust away. He constructs beautiful monuments and statues to see them crumble to the ground.

A common illustration of the Second Law of Thermodynamics compares the universe to a clock slowly running down. Even in this illustration the evolutionist is hard-pressed to identify who wound up the universe in the first place.

Needless to say, the theory of evolution directly contradicts the Second Law of Thermodynamics. Evolution leads to a decrease in randomness—that is, an increase in order—whereas the Second Law of Thermodynamics says the opposite should be occurring.

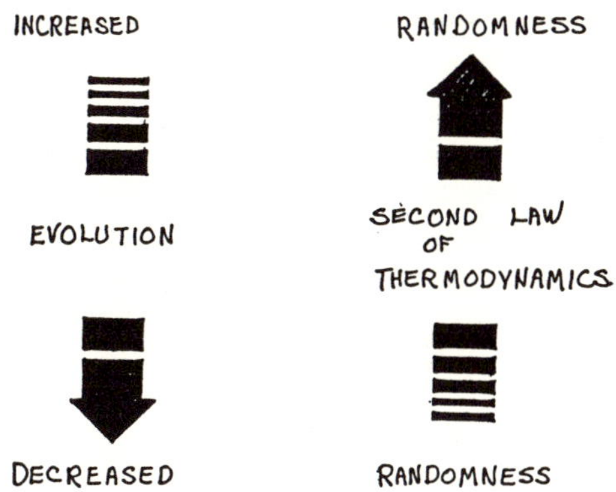

Thus the concepts of the creation and the curse harmonize perfectly with the universally accepted Second Law of Thermodynamics, but the theory of evolution stands in diametric opposition to this law.

The Second Law of Thermodynamics also exposes a fatal flaw in the uniformitarian concept. According to uniformitarianism, the present is the key to the past. (See Chapter 6.) The Second Law of Thermodynamics is in effect today; therefore, according to the uniformitarian theory, it must have been in effect forever. If this be true, the universe as we know it today would have completely lost all of its energy sources.

This theory does not imply that energy would have been destroyed, but rather that it would have dissipated to a level of complete randomness. If this forever existed, then the entire universe would be in a state of complete randomness and would lack useful energy. This is certainly not the condition of the universe today, however. Just a casual look at the starry sky on a cloudless night will provide more than sufficient evidence that high energy sources and order remain in the universe. Therefore, the Second Law of Thermodynamics could not have been operative forever.

At some point after the formation (creation) of the universe, it became effective. The uniformitarianist has no explanation for this; yet the creation concept easily resolves the dilemma. God created a perfect universe. Adam sinned, and because of sin God placed the entire creation under a curse. The result was degeneracy, or, in other words, the Second Law of Thermodynamics.

A so-called fact implied by the theory of evolution becomes fiction when viewed in the light of Scripture and the true laws of science.

Degeneracy of Life

According to the theory of evolution, man is progressively evolving toward a higher level of life. In other words, man is physically, mentally, spiritually, and emotionally improving. Yet, when we observe the world around us, we readily see that the very opposite is occurring. Man is actually degenerating and deteriorating. This scientific fact accords perfectly with the Second Law of Thermodynamics as well as the biblical teaching of the curse. We will discuss three examples of this degeneracy in man.

Prehistoric Man

Evolutionists presume that modern man evolved from some creature in the prehistoric past that had a much smaller brain cavity than man has today. (See Chapter 4.) They theorize that as eons passed, this brain cavity became progressively larger. If man truly did evolve, this gradual increase would seem not only logical but also essential; yet amazingly the fossils of modern man's so-called "nearest ancestors" actually had larger brain cavities than those of today. Both the Neanderthal Man and the Cro-Magnon Man had larger brain cavities than modern man, yet modern man supposedly evolved from them. Instead of proving evolution, the fossil record of early man strongly indicates degeneracy—the opposite of evolution.

Vestigial Organs

If evolution were occurring in man, a large number of organs in the human body should have no function, since some organs should be evolving for future use while other organs should be losing their function as new and improved ones replace them. Evolutionists in the early 1900's recognized that they could develop a strong argument for evolution if they could find vestigial (or rudimentary) organs in the human body. Conversely, they also recognized the strong argument against evolution if they identified no vestigial organs, or even just a small number of them. Therefore, they conducted a large-scale search to locate and number the vestigial organs in man.

After only a short time, evolutionists boldly proclaimed they had found 180 vestigial organs in man. They treated this as definite proof of man's past and present evolution. The argument was so convincing that even some university textbooks began to use it.[1]

For a long time, the concept of vestigial organs was one of the strongest arguments supporting the theory of evolution; however, as time passed, researchers discovered that many so-called vestigial organs are actually hormone producers. Furthermore, many others are normally inactive, but become active when the body is subjected to stress. The latest research indicates that there are very few or, according to some scientists, no vestigial organs.

Until recently, evolutionists commonly pointed to the tonsils as a prime example of a vestigial organ. Today we know that the tonsils are significant masses of lymphoid tissue which function in capturing foreign particles, especially bacteria.

The most commonly used example of a vestigial organ at present is the appendix. There is no known function for the appendix in man, and it is well known as a "seat of disease." Evolutionists argue that the appendix is considerably larger in many other animals and aids digestion of coarse foods such as cellulose.

Evolutionists Dodson and Dodson present the typical anti-creationist view: "The appendix of man is easily understandable as a degenerating legacy from ancestors with a much coarser diet, but it is inexplicable why a useless and disease-ridden structure would have been created especially to plague him."[2] What Dodson and Dodson do not realize is that man was not created in his present form. God created man in perfection, but due to sin and the curse he has degenerated from perfection to the imperfect disease-ridden creature that he is today. Furthermore, if modern man would include more fiber in his diet, he would not get appendicitis.

One problem evolutionists face concerning the appendix is that many monkeys have no appendix.[3] Modern man supposedly evolved from an ancestor similar to the monkey, yet the absence of the appendix in many monkeys would indicate they have evolved further than man. Thus the argument in favor of evolution based on a vestigial appendix is very superficial.

Evolutionists have gone to great lengths to prove their theory by maintaining that there are a few vestigial organs in the human body, yet they cannot point to one organ that is in the process of evolving into something useful for future man. Let us emphasize once again that the degeneration of man could produce vestigial organs, harmonizing perfectly with the Second Law of Thermodynamics and the curse but contradicting the concept of evolution.

Again, the "facts" for evolution have become fiction, and in this case the evidence for evolution has become the evidence against evolution.

Genetic Diseases

Another example of degeneracy within man is genetic diseases, such as albinism, sickle-cell anemia, hemophilia, mongolism, Klinefelter's syndrome, and Turner's syndrome.

Scientists have found that man's genetic makeup is becoming contaminated or, in other words, degenerating rather than improving or evolving. Hickman and Hickman describe the widespread incidence of genetic defects:

> *Geneticists estimate that one person in every four or five of the population is born with a defective gene, which can bring about serious consequences to the population, depending on whether the genes are dominant or recessive. It is thought that one conception in every five results in a genetic defect that destroys the individual at some time before maturity.*[4]

From these brief examples we see that man is degenerating. How then can man at the same time be improving or evolving?

The Wives of Cain and Seth

At this point, it is interesting to discuss a question Christians often face: "Where did Cain and Seth get their wives?" The answer is simple; they married their sisters. Genesis 5:4 states that Adam had sons and daughters.

The idea of Cain and Seth marrying their close kin may seem genetically unhealthy, but we must remember that their parents, Adam and Eve, were genetically pure. If Adam and Eve had not sinned, there would have been no curse of degeneration, and their offspring for infinite generations could have intermarried without yielding any harmful genetic side effects.

Today, after thousands of years of degeneration, the genetic impurity of man has increased to such an extent that if close relatives marry, their genetic combination magnifies the probability of genetic defects. This is one reason why it is now unlawful to intermarry within a family. This restriction did not apply in Adam's time when genetic diseases were nonexistent.

The World Before the Flood

Obviously, no one alive today has ever seen the world that existed before the Flood. Fortunately, Scripture does provide a few glimpses of this world that overflowed with water and perished. (See II Peter 3:6.) Combining Scripture with the fossil record provides a fairly accurate picture of the world that Noah lived in before the Flood.

It must have been a beautiful world, for it had a vapor canopy that produced a year-round tropical climate and a worldwide array of vegetation. This vegetation, in turn, became both the food and the abode of an amazing group of animals called "dinosaurs."

Vapor Canopy

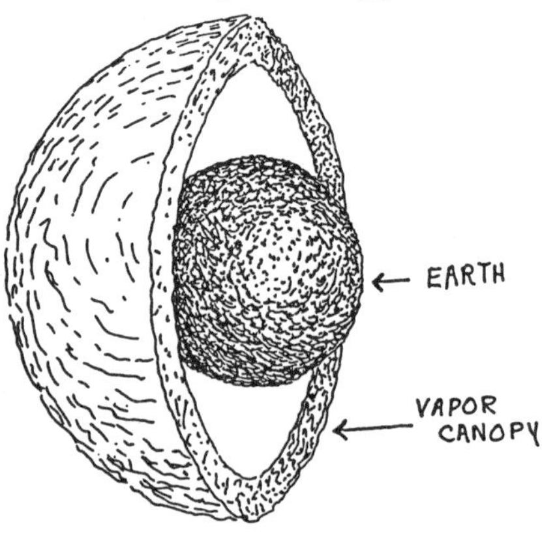

"And God said, Let there be a firmament in the midst of the waters, and let it divide the waters from the waters. And God made the firmament, and divided the waters which were under the firmament from the waters which were above the firmament: and it was so" (Genesis 1:6-7).

This passage of Scripture describes the second day of creation, in which God created a vapor canopy that completely enveloped the earth. This canopy surrounded the earth until it broke apart to produce rain for forty days and forty nights (Genesis 7:11-12).

The vapor canopy was more than the atmospheric cloud cover of today, "For the LORD God had not caused it to rain upon the earth" (Genesis 2:5). If the vapor canopy was merely the same cloud cover that exists today, then rain showers and thunderstorms would have watered the earth. Rather, a different more efficient system was in effect. "There went up a mist from the earth, and watered the whole face of the ground" (Genesis 2:6).

We should note that the mist watered the entire face of the ground, that is, every square inch of the earth. Consequently, there were no droughts, no deserts, and no wastelands. Today, the land suitable for habitation and cultivation comprises less than one half of the total land area of the earth.

Genesis 1:9 provides another scriptural reason to believe that the pre-Noachian vapor canopy was different from the cloud cover of today: "And God said, Let the waters under the heaven be gathered together unto one place, and let the dry land appear: and it was so." God specifically gathered the waters under the heaven together in one place, whereas before they had totally encircled the globe just as the waters above the firmament (Genesis 1:7).

Significantly, this passage does not state that God gathered together the atmospheric water into clouds, leaving relatively dry expanses of open atmosphere as is the case today. Apparently, he left the atmospheric water uniformly distributed as a vapor canopy.

We should note that our present atmosphere does not contain enough water to produce a global rain lasting forty days and forty nights. According to scientific estimates, if all the water in our present atmosphere precipitated out and spread evenly over the global surface, it would produce a layer of water perhaps two inches in depth. This is another strong indication that the water above the firmament in Genesis 1:7 was not merely the cloud cover of today, but a vapor canopy containing many times the amount of water in our present atmosphere.

The Bible records the existence of seasons immediately after the Flood. "While the earth remaineth, seedtime and harvest, and cold and heat, and summer and winter, and day and night shall not cease" (Genesis 8:22). Although Genesis 1:14 mentioned seasons, there is no scriptural account of their taking effect until after the Flood. We should also note that Adam and Eve were both naked (Genesis 2:25) and would have remained in this condition if they had not sinned. All this indicates that the climate must have been continually warm and comfortable before the Flood.

Is this all an incredible coincidence, or does the concept of a vapor canopy fit in with the total picture that the Bible paints of the beginning? In the beginning, God created a world of beauty and perfection. This world had a perfect climate, a perfect environment, a perfect variety of animals to inhabit the environment, a perfect variety of plants to serve as food for animals and for man, and a perfect man to exercise rule and dominion over it all. Then entered sin, the fall, and the curse. From that day forward degeneration became the ruling principle not only of man but also of the entire earth. A worldwide flood eventually destroyed the perfect climate of the earth, resulting in the erratic climate of today.

If the Bible is not the inspired Word of God and is not totally accurate, then how could the Old Testament record the effects of a vapor canopy in earth's atmosphere, such as even temperatures, no wind, and no rain? When Moses recorded the Genesis account, man simply did not know what produced seasons and temperature variations on the earth.

Evolutionists sometimes ridicule the idea of a pre-Noachian vapor canopy, without realizing their own theory proposes the same idea. Most evolutionary theories of the earth's formation postulate that the earth was, at one period of its evolutionary history, a hot molten mass. As the molten mass began to cool, it formed a thin crust, a mantle, and a core. Volcanic action created water droplets high in the atmosphere. Before these droplets hit the super-heated surface of the earth, they vaporized again, due to the intense heat. According to evolutionary theory, then, a very thick vapor canopy once existed.

Evolutionist Dickerson describes the process as he envisions it:

> *The earth became stratified into a core, a mantle and a crust as a result of the heat released as the planet was built up by accretion. The orginal surface may have been too hot for water to remain liquid, but as soon as the temperature had fallen below the boiling point water released from the interior by out-gassing processes such as volcanism would have condensed to form the original oceans.*[5]

Dodson and Dodson also comment on atmospheric water vapor:

> *A large quantity of the hydrogen and oxygen were probably united as water, but for long ages it was certainly present only as superheated steam. The hot vapors would rise toward the cold outer layers of the atmosphere, condense and fall as rain, only to be convert-*

> *ed to steam again before striking the earth. Gradually the earth cooled sufficiently to permit the rainfall to strike the earth, then to begin to form pools and larger bodies of water.*[6]

Creationists do not necessarily agree with these evolutionary concepts, but the above statements demonstrate that the vapor canopy ridiculed by evolutionists is a part of their own theory.

In summary, the vapor canopy hypothesis accounts for the following facts: (1) the description of the waters above the firmament (Genesis 1:6); (2) the initial absence of rain (Genesis 2:5-6); (3) the immense quantity of water required for a rain of forty days and forty nights; (4) the temperate climate, vastly different from that of today, required for the ferns, lush vegetation, and dinosaurs of the fossil record; and (5) the long life spans of people before the Flood.

Longevity of the Patriarchs and the Ozone Layer

One of the most amazing phenomena of the Old Testament is the long lives of men before the Flood. For example, Adam lived 930 years, Seth 921 years, Enos 905 years, Jared 962 years, and Methuselah 969 years. After the Flood, the Bible chronology shows a gradual shortening of the life span of the various patriarchs. As time passed, the expected life span continued to decline until it became seventy years (Psalm 90:10).

Below is a graph showing the decline in age from Adam to modern man.

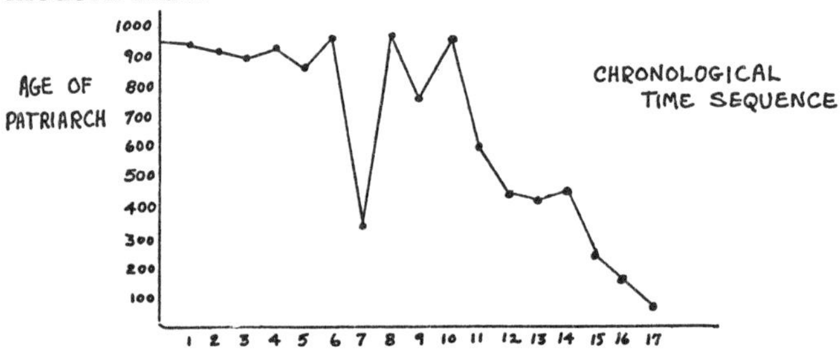

Why did men attain such great ages before the Flood and why did the average life span dwindle to seventy years? The most reasonable answer is that a vapor canopy and a thick ozone layer existed before the Flood in contrast to no vapor canopy and an extremely thin ozone layer today.

Ozone is a shield against solar radiation, which many believe to be the primary factor in aging. Ozone originates from short wave solar radiation striking O_2 (oxygen) molecules in the upper atmosphere and converting them to O_3 [ozone] molecules. These O_3 molecules can break down into O_2 molecules again by long wave radiation. Long wave radiation comes from solar radiation striking the earth and reflecting back into the atmosphere.

Water vapor can stop long wave radiation. In the antediluvian age, just as today, solar radiation struck O_2 molecules and converted them to O_3 molecules. The solar radiation that the ozone layer did not absorb continued to travel through the atmosphere until it struck the earth's surface and reflected back in the form of long wave radiation. Before the long wave radiation could reach the ozone layer to break it down (as occurs today), in the antediluvian age the vapor canopy, which was at a much lower altitude than the ozone layer, reflected it back to earth.

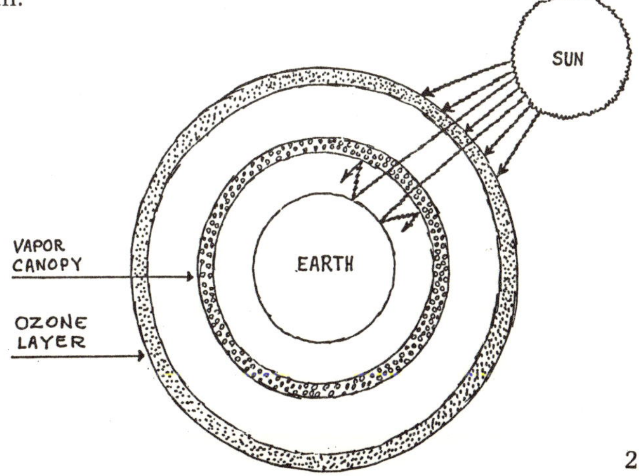

With the dissipation of the water vapor canopy after the Flood, the long wave radiation could reach the ozone layer from the earth and begin its deteriorating effect upon the ozone layer. A number of years passed before an equilibrium was established between the formation of O_3 and its breakdown again into O_2 by long wave radiation. As the ozone layer became thinner and thinner, its capability to shield against solar radiation became less and less. Perhaps this caused the gradual transition from long life spans to those of today.

There is a bit of a paradox in the nature of ozone today, for it is both a blessing and a curse. On the one hand, it is beneficial to life because it filters out deadly solar radiation in the upper atmosphere. In fact, without the ozone layer, life could not exist on the earth. On the other hand, ozone is detrimental because it is a deadly poison. The winds in the upper atmosphere often carry ozone down to low altitudes. Plants and animals then ingest the ozone (although the winds carrying it dilute it somewhat). This has an adverse effect on both plants and animals and often results in premature death.

The conditions of the atmosphere and the ozone layer were totally different before the Flood. With the vapor canopy there were no climatic variations; thus no winds were formed. The ozone remained in the upper atmosphere and did not adversely affect animals or plants. At the same time, it was a constant shield against solar radiation.

We conclude that the combination of the vapor canopy and the ozone layer in filtering out solar radiation probably accounted for the longevity of the patriarchs before the Flood.

Tropical Climate

The fossil record testifies to the existence of a warm tropical climate ranging across the entire globe before the Flood. Fossils found from the Arctic Circle to Antarctica indicate that a greenhouse-type environment must have existed throughout the whole earth.

Interestingly, evolutionists have no alternative explanation for a wide-ranging tropical climate in the prehistoric past, but the evidence accords perfectly with creationism. Briefly restated, many creationists believe a vapor canopy once existed and produced a greenhouse effect, resulting in a warm, moist, uniform climate. This climate produced lush vegetation, which in turn became the food and abode of an amazing variety of warmth-loving animals.

Today, fern trees and lush tropical vegetation grow only in the equatorial latitudes. This was not the case, however, throughout earth's history. As the following quotations from scientific literature demonstrate, the earth was once a universal greenhouse.

Elso Barghoorn describes an earth with little variation in climate:

> *From the paleontologic evidence available, it would appear that there was very slight climatic zonation between high and low latitudes during the major part of the Carboniferous. Very similar floras occur in regions that today show extreme climatic differences; for example, northeastern Europe and the eastern Mediterranean, or Nova Scotia and central Kansas.*[7]

Edwin Colbert also discusses the uniform climates of earth in the past:

> *As in the earlier period the climates of the earth were in general rather uniform. Tropical floras and the animals that one might expect in such associations of plants lived at fairly high latitudes—as far north as 45°.*[8]

As far as past climates can be interpreted from the record of fossil vertebrates, it would appear that during much of earth's history the world has enjoyed uniformly warm, equable climates over most of its surface. There have been digressions from this pattern of uniformity at times, and various complications in it, but the general picture of past vertebrate life is that of warmth-loving animals living over wide ranges of latitude, from the southern tips of the continental land masses through the middle latitudes to regions as far north as the Arctic Circle.[9]

The following quotations further document the scientific evidence for a universal warm, humid climate with widespread distribution of tropical vegetation:

> *In Eocene and during much of Oligocene, most of the earth enjoyed warm and humid climates.*[10]
>
> *Fossils deposited at that time indicate that the climate was much milder than at present, for the fossil record shows that alligators, sassafras trees, and magnolias were more or less continuously distributed from southeastern United States to eastern China, from the banks of the Yangtse to the banks of the Suwannee.*[11]
>
> *Climates must have been warm and moist throughout a great part of Carboniferous time to support the abundant vegetation of ancient club mosses, lepidodendrons, horsetails, and ferns that constituted the forest of those distant days.*[12]
>
> *Palm trees grew all over the U.S., and New York State was as warm as present day Florida.*[13]
>
> *For example, Plate I illustrates a specimen of a palm and one of a cycad that were collected from the Eocene rocks in Kupreanof Island in southeastern Alaska. These leaves were associated with the remains of laurels, magnolias, acacias, peppers, and other forms which clearly indicate conditions much warmer than exist in that region today; in fact, they were subtropical.*[14]
>
> *Fossil palms and magnolias about 70,000,000 years old found in Greenland indicate a warm, moist climate at that time.*[15]
>
> *Jurassic floras, botanically rich in varied types, are known from some of the most northerly land areas of the Arctic, such as northern Greenland and Franz Josef Land, north of the 80th parallel. The most remarkable*

evidence of the non-glacial climate of polar regions in Jurassic time, however, was discovered on the Third Byrd Expedition to Antarctica. Fossil plants and coal were collected on Mount Weaver, at an elevation of 10,000 feet at latitude 86°58′ S. The plants themselves, interestingly enough, belong to genera known from Jurassic sediments of temperate and even Arctic regions of the Northern Hemisphere. The development of coal-forming swamps close to the South Pole demands an explanation allowing for extraordinary climatic change. Botanically also, Jurassic floras present the curious anomaly of a well-differentiated vegetation developing under the physiologic conditions of the polar night.[16]

Extensive coal deposits, even from such unlikely places as Antarctica and Spitsbergen, are generally interpreted as having accumulated during a humid, temperate to subtropical period.[17]

In the late 1960's, fossil amphibians and reptiles were discovered in 200 million year old coal swamps in Antarctica.[18]

Truly warm temperate forests, dominated by such forms as dawn redwoods, maples, beeches, oaks, sycamores, and basswoods, are widely recorded in rocks of this age in the belt extending from Central Alaska to West-Central Greenland and eastward to Spitsbergen and Siberia (Chaney 1947). In Greenland the fossil-leaf beds occur on the bleak Arctic wastes of Disco Island, almost within sight of the glacial mantle of ice covering the mainland.[19]

The luxuriant growth of broad-leaf hardwood forest in high Arctic latitudes persisted from the Cretaceous into the Eocene and probably the Oligocene, indicating a prolonged continuation of humid warm temperate, or at least temperate forest climate in polar regions. Evidence for this may be found in both Arctic and Ant-

> *arctic regions.* [20]
>
> *The marine reptiles of the Cretaceous period, notably the ichthyosaurs, plesiosaurs, and mosasaurs, were as widely spread across the face of the earth as the land-living forms—an indication that warm seas extended far into the northern and southern latitudes of the globe.*[21]

Another interesting indicator of warm humid climates in the past is the discovery of frozen mammoths in Alaska and Siberia. The remains of the creatures have been found literally by the thousands. Many had undigested grasses, bluebells, buttercups, tender sedges and wild beans still in their stomachs and mouths. Many have remained frozen for thousands of years in the position which they fell. Obviously, these animals were grazing in a warm humid climate when a sudden, cataclysmic event froze the mammoths instantly. This is also true for many other animals that have been found quick-frozen, such as sheep, camels, rhinoceroses, and bison.[22]

Desert areas of today, such as the Sahara, the American West, the Great Australian Desert, and the Chilean Atacama, also yield evidence of past tropical vegetation and animals, indicating that there were few or no desert regions in the prehistoric past.

The documentation of worldwide tropical, or at least semitropical, conditions could go on and on. Suffice it to say that fossil records supply strong evidence of a warm humid climate on the earth in the prehistoric past. Again evolutionists do not have a reasonable explanation, yet this accords perfectly with the "greenhouse" effect of the vapor canopy.

Growth Rings in Trees

In temperate zones, as well as in some tropical regions that have marked wet and dry seasons, climatic conditions are not suitable for the continued growth of woody trees and shrubs throughout the entire year. Consequently, the outer growth tissue, called the cambium, experiences alternating periods of activity and dormancy. This alternation results in annual growth rings or growth layers. As Barghoorn explains, "In climates with distinct seasons, seasonal effect is almost invariably reflected in pronounced annual growth rings."[23]

If the climate before the Flood was warm, humid, and uniform, then one of the strongest indicators would be the absence of annual growth rings in the fossil remains of at least some woody plants. This is actually the case: "One impressive indicator of uniform climate over great areas of the carboniferous continents is the general absence of annual growth rings in coal-swamp trees."[24]

Although evolutionists have no satisfactory explanation, they have to acknowledge a broad-ranging tropical or semitropical climate in earth's past. Of course, this contradicts the concept of uniformitarianism, for if the present is truly the "key" to the past, seasonal variation should have existed in the past just as it exists today.

Dinosaurs

Fossil records indicate that many years ago colossal creatures called dinosaurs roamed the earth. Many weighed perhaps as much as one hundred tons,[25] approximately fifteen times the weight of an adult African bull elephant. They apparently ranged almost from pole to pole, for their remains have been found in abundance on all continents except Antarctica. Dinosaur relics have been found as far south as southern Argentina (latitude 48ºS), and their footprints occur on the island of Spitsbergen, well inside the Arctic Circle (near latitude 80ºN).[26]

Many questions about the dinosaurs remain unanswered, as John Ostrum notes:

> *Many intriguing questions about dinosaurs remain. Why were the dinosaurs so unlike any other animals living now or in the past? How did they live? Why were so many of them so large—some unbelievably huge? Did they simply never stop growing? How long did it take for them to grow so large? How could such creatures move about? How could they have eaten enough to keep them going? We don't know. And finally, the greatest mystery of all: the vast array of dinosaurian kinds all became extinct apparently quite suddenly, about 65 million years ago. After dominating the earth for 140 million years, they disappeared. Why? What could have killed so many different kinds of well-adapted and highly successful creatures?*[27]

Seemingly one of the greatest puzzles of the ages is, "What happened to the dinosaurs?" Theodosius Dobzhansky remarks, "Just what caused the dying out of these varied and apparently well-adapted animals at the end of the Cretaceous period is obscure."[28]

James Michener comments "But why did they and the little mammal live when the great reptiles vanished? This remains one of the world's supreme mysteries."[29]

To complicate this puzzle, many other life forms became extinct at the same time, as Ostrum observes:

> *At the same time that the dinosaurs were dying out, a dramatic extermination of other life forms occurred. In the sea the reptilian monsters disappeared, together with the coil-shelled ammonites, relatives of the modern squid and nautilus. Massive extinctions occurred in the sea's microscopic planktonic life. On land certain plants vanished and the flying reptiles disappeared.*[30]

Some scientists have suggested that disease caused the extinction of the dinosaurs, but Roger Batten explains why this could not be the case:

> *Some students have suggested that a pathogenic fungus or some disease wiped them out, but if we take diseases of today as a model, they usually affect only a single species and there is no known species that has been eradicated completely, even by a pandemic. Another theory was considered that mammals might have eaten the dinosaur eggs, but again it is difficult to believe this would have been 100 percent effective.*[31]

Evolutionist Nicholas Hotton III concedes that there must have been a "universal factor" causing the extinction of the great variety of organisms, but he quickly admits he can scarcely guess at the factor.[32] One thing is certain: evolutionists do not want to link the extinction of the dinosaurs and other life forms with a catastrophic cause such as the Flood. Hotton acknowledges as much: "What really did them in is not known, but without strong supporting evidence we should not explain extinction by single or catastrophic causes, or by causes outside of the biologic world."[33] Yet, the environmental changes caused by the Flood provide the most reasonable scientific explanation for the extinction of the large variety of life forms including dinosaurs.

Another puzzle to scientists is how the dinosaurs found

enough food. When we consider that an adult bull elephant spends eighteen of every twenty-four hours searching for and consuming 300 to 600 pounds of fodder, we easily see why many scientists are puzzled about the dinosaurs, some of which were fifteen times the weight of the elephant. The answer rests in the large variety of lush vegetation in the antediluvian world due to the ideal environment. Most dinosaurs were, in fact, plant eaters:

> *The fact that most dinosaurs were plant eaters also certainly contributed to their over-all success, for it gave them access to virtually every niche commensurate with their size range (chicken to whale size) throughout the Jurassic and Cretaceous periods."*[34]
>
> *The largest dinosaurs (Diplodocus and Brachiosaurus, the latter estimated to have weighed some 50 tons) were, however, vegetarians.*[35]

Any factor harming the plant life consumed by the dinosaurs would certainly have had an immediate detrimental effect upon them. Dott and Batten identify such a disruption of the food chain as a probable cause of extinctions:

> *The most likely cause of extinctions is the most difficult to substantiate. The most delicate balance in nature involves the ecosystem; that is, the interaction of organisms in the biological and physical structure of the community. The success of any community is largely dependent on a few key species of plants and/or animals in the food chain. If any of these species are removed, the effect will be felt immediately through the whole food pyramid of the community—a sort of domino effect.*[36]

In all probability, the harsh new environment after the Flood was not conducive to widespread lush vegetation. As a result, the dinosaurs did not have an adequate food supply and simply starved to extinction.

Was There Room On The Ark For Dinosaurs?

In order to determine whether the ark had room for all the animals, we must first determine the capacity of the ark. According to Genesis 6:14-16, the ark was three hundred cubits long, fifty cubits wide and thirty cubits high. It had three stories, and each story was divided into rooms.

It is impossible to establish a certain, uniform measurement for the Old Testament cubit, but the smallest estimate is 17.5 inches. Using this minimum figure, the ark had a total deck area of 95,700 square feet, slightly more area than twenty college basketball courts.[37] It had a minimum total volume of 1,400,461 cubic feet or in other words the capacity of a least 522 railroad boxcars.[38]

According to renowned taxonomist Ernst Mayr, over a million species of animals exist today.[39] Since only those animals who breathed air through nostrils entered the ark (Genesis 6:17; 7:21-22), we can eliminate the aquatic species from our calculations. The figure then dwindles to 17,500 species.[40] The average size of these 17,500 species is approximately that of a German shepherd dog. Allowing one male and one female for each species, the ark needed room for 35,000 animals the size of a German shepherd dog. This translates into approximately 150 boxcars.

Some people contend that the insects were in the ark, although the Bible does not specifically so state. For the sake of argument, we shall include them in our calculations. According to Mayr, there are presently 850,000 species of insects.[41] Multiplying 850,000 by two and allowing four inches of crawling and flying space for each insect, we conclude that all of the insects needed approximately fifty boxcars.

Allowing 150 boxcars for breathing animals and fifty boxcars for insects, we get a total of two hundred boxcars, which is still less than half the 522-boxcar minimum capacity of the ark. The excess capacity would have provided ample space for all extinct species as well as room for Noah, his family, and food. Despite the great size of some dinosaurs such as *Brontosaurus* and *Tyrannosauras rex,* most extinct vertebrate animals were very small, ranging from the sizes of mice to steer.[42] These calculations indicate that all extinct breathing animals could have fit in less than one-third of the ark's volume.

In conclusion, the ark had ample space for both living and extinct animals. There was also ample room for Noah, his family, and food storage, as well as space for relaxation and recreation.

Concerning dinosaurs, if they were still alive at the time of the Flood and if they survived the Flood, then they perished in the postdiluvian world due to the drastic changes in climate and vegetation. Even so, there was still ample room for them in the ark.

Dinosaurs and the Bible

Since dinosaurs actually existed, why do we find no biblical comment on these amazing creatures? Perhaps the Bible does comment on dinosaurs. Job 40:15-24 describes the "behemoth":

> *"Behold now behemoth, which I made with thee; he eateth grass as an ox. Lo now, his strength is in his loins, and his force is in the navel of his belly. He moveth his tail like a cedar: the sinews of his stones are wrapped together. His bones are as strong pieces of brass; his bones are like bars of iron. He is the chief of the ways of God: he that made him can make his sword to approach unto him. Surely the mountains bring him forth food, where all the beasts of the field play. He lieth under the shady trees, in the covert of the reed, and fens. The shady trees cover him with their shadow; the willows of the brook compass him about. Behold, he drinketh up a river and hasteth not: he trusteth that he can draw up Jordan into his mouth. He taketh it with his eyes:*

his nose pierceth through snares."

Scientists sometimes refer to the dinosaur by the term behemoth, as Dodson and Dodson do in their well-known college textbook *Evolution, Process and Product*: "Upright posture does not relate merely to supporting the bulk of behemoths: the small, early dinosaurs were completely upright, and similar types continued throughout the Mesozoic."[43]

Some reference books identify the behemoth of Job 40 with the elephant or perhaps the hippopotamus, but verse 17 apparently describes an animal with a much larger tail than either of them: "He moveth his tail like a cedar." It would be hard to imagine the tiny tail of an elephant or hippopotamus moving like a cedar. Although no one can establish with certainty what this animal was, it is possible that Job 40 describes the dinosaur.

How Did the Animals Get to the Ark?

A previous section demonstrated that the ark provided ample space for two of every species of breathing animal. Now we discuss how these animals got to the ark. Our only scriptural evidence is Genesis 6:20: "Two of every sort shall come unto thee." Contrary to the thinking of many critics, Noah did not have to go out and catch every kind of animal. Rather, God sent the animals to Noah. God could have used several ways to accomplish this; we shall speculate on a few possibilities.

Apparently, the animals did not have to travel through hot deserts or cold regions since a warm tropical climate prevailed over the entire earth before the Flood. Given the uniform climate, the various animal species probably had an even distribution. In other words, due to even climatic conditions and ample food supplies, the same "climate community" involving every kind of animal existed throughout the entire land surface of the globe, including Mesopotamia, where Noah and the ark were. If so, the animals did not have far to travel, and even slow-moving animals, such as sloths and turtles, could have easily reached the ark in time.

Another possibility is that God started the migration of animals toward the ark when He commanded Noah to build the ark. If this is true, the time factor was not critical, for the animals had 120 years to reach the ark (Genesis 6:3). Presumably animals, as well as man, lived much longer before the Flood than they do today. Perhaps God caused entire herds of animals to migrate toward the ark. Several generations could have lived and died before the herd reached the ark, at which time God could have chosen the fittest male and female to enter the ark.

Perhaps it was a combination of these two possibilities. Whatever the case, one thing is certain: they came!

Hibernation

Before the Flood, the earth had almost a perfect climate. The temperature remained basically constant. The soil was extremely fertile due to the upward leaching of soil nutrients. There was ample water either from artesian wells or from the mist that went up from the earth (Genesis 2:6, 10-14). Under these conditions, animals certainly had no need for hibernation.

When did animals start hibernating? Scientists know what causes animals to hibernate. Adverse climatic factors such as cold, drought, lack of food, and extreme heat bring on hibernation (winter) or aestivation (summer). Evolutionary scientists still face the dilemma, however, as to when and how animals first began to hibernate.

Since there are not direct scriptural references to hibernation, we cannot be dogmatic on the subject. Nevertheless, the Noachin account would provide a reasonable explanation. To investigate this possibility, let us look at Genesis 6:21. There God told Noah, "Take thou unto thee of all food that is eaten, and thou shalt gather it to thee; and it shall be for food for thee, and for them [the animals]." This verse does not specifically state that all the animals on the ark would be eating; God simply told Noah to gather all the food necessary. Since hibernating animals do not eat food, this verse would not apply to them. Only those animals incapable of hibernating on the ark would have required food, and God commanded Noah to gather food to sustain them.

Hibernation would have also provided a good solution to the problem of waste disposal. Only animals not in hibernation would have produced excrement. Noah and his family could have easily disposed of the wastes of non-hibernating animals.

Footnotes
Chapter One

[1]Carl Dunbar and Waage, *Historical Geology* (John Wiley and Sons, 1969), p. 111.

[2]Edward O. Dodson and Peter Dodson, *Evolution, Process and Product* (D. Van Nostrand Co., 1976), p. 49.

[3]Thomas F. Heinze, *Creation Vs. Evolution* (Grand Rapids: Baker Book House), p. 22.

[4]Cleveland P. Hickman and Cleveland P. Hickman, Jr., *Biology of Animals* (C. V. Mosby Co., 1972), p. 307.

[5]Richard E. Dickerson, "Chemical Evolution and the Origin of Life," *Scientific American*, Vol. 239, No. 3 (Sept. 1978), p. 73.

[6]Dodson and Dodson, pp. 109 and 110.

[7]Elso S. Barghoorn, *Evidence of Climatic Changes as Revealed by Vertebrate Paleoecology*, ed. Harlow Shapley (Harvard University Press, 1953), p. 257.

[8]Edwin H. Colbert, *The Record of Climatic Changes as Revealed by Vertebrate Paleoecology*, ed. Harlow Shapley (Harvard University Press, 1953), p. 257.

[9]Colbert, p. 269.

[10]Theodosius Dobzhansky, *Evolution, Genetics and Man* (John Wiley and Sons, 1955), p. 304.

[11]David J. Merrell, *Evolution and Genetics* (Holt, Rinehart & Winston, 1962), p. 73.

[12]Colbert, p. 256.

[13]Ernst J. Opik, "Climate and the Changing Sun," *Scientific American* (June 1958), p. 4.

[14]Richard M. Field, *Geology*, 4th ed. (Barnes & Noble, 1934), p. 343.

[15]*The World Book Encyclopedia*, (Field Enterprises Educational Corp., 1960), VI, 367.

[16]Barghoorn, p. 243.

[17]Erling Dorf, "Climatic Changes of the Past and Present," *American Scientist*, Vol. 48, No. 3 (Sept. 1960), p. 345.

[18]Dodson and Dodson, p. 21.

[19]Dorf, p. 347.

[20]Barghoorn, p. 244.

[21]Colbert, p. 264.

[22]Donald W. Patten, *The Biblical Flood and the Ice Epoch* (Pacific Meridian Publishing Co., 1966), p. 104-109.

[23]Barghoorn, p. 242.

[24]*Ibid.*

[25]John H. Ostrom, "A New Look at Dinosaurs," *National Geographic*, Vol. 154, No. 2 (Aug. 1978), p. 171.

[26]*Ibid.*, p. 170.

[27]*Ibid.*, p. 167.

[28]Dobzhansky, p. 292.

[29]James A. Michener, "Life and Death of the Dinosaur," *Reader's Digest*, Vol. 105, No. 631 (Nov. 1974), pp. 233-234.

[30]Ostrom, p. 178.

[31]Robert H. Dott, Jr. and Roger L. Batten, *Evolution of the Earth* (McGraw-Hill Book Co., 1971), p. 581.

[32]Nicholas Hotton, III, *The Evidences of Evolution of the Earth* (American Heritage Publishing Co., 1968), p. 100.

[33]Hotton, p. 99.

[34]*Ibid.*, p. 98.

[35]Dobzhansky, p. 291.

[36]Dott and Batten, p. 582.

[37]Hotton, p. 60.

[38]John C. Whitcomb, Jr. and Henry C. Orris, *The Genesis Flood* (The Presbyterian and Reformed Publishing Company, 1961), p. 10.

[39]Kelly L. Segraves, *The Great Dinosaur Mistake* (Kelly L. Segraves Beta Books, 1975), p. 16.

[40]*Ibid.*, p. 12.

[41]*Ibid.*, p. 14.

[42]Dunbar and Waage, p. 391.

[43]Dodson, p. 218.

Chapter 2
THE FLOOD

> *"And, behold, I, even I, do bring a flood of waters upon the earth, to destroy all flesh, wherein is the breath of life, from under heaven; and every thing that is in the earth shall die"* (Genesis 6:17).

With the exception of creation, the greatest geological event in the planet earth's history was the Great Flood described in Genesis 6-8. Most people have the impression, however, that the Flood was merely a gentle raising of the water level over the earth, having little or no effect upon the earth's surface. Perhaps we remember simple descriptions of the Flood from childhood Bible stories and never stop to analyze the drastic physical consequences of such a flood.

Some people suggest that perhaps the Flood was not universal in extent, but covered only the Mesopotamian region. As we shall see, however, the Flood was not only universal in extent but also cataclysmic in effect.

Scriptural Evidence of a Worldwide Flood

At least thirty references in Genesis alone describe the Flood as universal. We shall cite several of these passages to establish this conclusion.

> **Genesis 6:13**—*"And God said unto Noah, the end of all flesh is come before me; for the earth is filled with violence through them; and, behold, I will destroy them with the earth."*

This passage indicates that God sent the Flood at least to all inhabited parts of earth.

> **Genesis 6:14**—*"Make thee an ark of gopher wood; rooms shalt thou make in the ark, and shalt pitch it within and without with pitch."*

There would have been no need for an ark if the Flood were only regional. God could have simply directed Noah to leave the region to be flooded.

> **Genesis 6:17**—*"And, behold, I, even I, do bring a flood of waters upon the earth, to destroy all flesh, wherein is the breath of life, from under heaven; and every thing that is in the earth shall die."*

The Flood destroyed all flesh with the breath of life from under heaven. The entire land surface of the globe is under heaven. If the Flood were not universal in scope, some animals in land areas under heaven would have lived through the Flood. (See also Genesis 7:4, 21-23.)

> **Genesis 6:19**—*"And of every living thing of all flesh, two of every sort shalt thou bring into the ark, to keep them alive with thee; they shall be male and female."*

If the Flood were not universal in extent, God could have simply caused the animals to migrate from the Mesopotamian region.

> **Genesis 7:19-20**—*"And the waters prevailed exceedingly upon the earth; and all the high hills, that were under the whole heaven, were covered. Fifteen cubits upward did the waters prevail; and the mountains were covered."*

If the Flood had been confined to the Mesopotamian region, then all the hills under heaven would not have been covered. Since all the mountains were under water and since water seeks its own level, in just a short time the water would have flowed over the entire face of the earth.

> **Genesis 9:11-13**—*"And I will establish my covenant with you; neither shall all flesh be cut off any more by the waters of a flood; neither shall there any more be a flood to destroy the earth. And God said, This is the token of the covenant which I make between me and you and every living creature that is with you, for*

perpetual generations: I do set my bow in the cloud, and it shall be for a token of a covenant between me and the earth."

If the Flood had been merely local in extent, centering around the Mesopotamian region, then God has not fulfilled His covenant with Noah, for there have been many local floods in various parts of the world since the time of Noah.

Flood Traditions

With some variations, almost every culture on the earth has myths and legends of a worldwide flood.[1] In fact, ancient people from all six continents and from forty different cultures have a flood tradition. Even isolated peoples such as the Hopi Indians in southwestern America and the Incas high in the Peruvian Andes have legends of a worldwide, devastating flood.

The many flood traditions found throughout the world are highly significant. We would expect that as people who heard a firsthand account of the Flood from Noah and his sons scattered across the earth, they would carry the story with them.[2] If we assume the biblical record to be correct, we can readily account for the general similarity among the widely scattered flood traditions as well as their variations.

This position is clearly stated in the words of Byron Nelson:

> *The assumption of the truth of the deluge account of Genesis offers a satisfactory explanation of the fact that the more ancient a tradition is, and the more close geographically to that central place in Asia where the Bible says the ark landed, the more reasonable and sane is the legend, the more generally like the biblical account, and vice versa. It is well-known that a story loses in accuracy each time it is told, and this accounts for the general situation as it exists. An incident related by Reverend John Inglis serves to illustrate why some leg-*

ends have become so different from the biblical narrative. A native of one of the islands of the New Hebrides was once listening to a missionary reading the biblical account of the Flood. In the midst of the reading the native said, "Stop. That is almost the same as what we have been told." He then related his tradition. "But," he added, "your fathers, having written an account for you, while ours only told it to their children, yours must be more correct than ours.[3]

In summary, the vast array of flood traditions from so many different cultures is strong evidence that a universal flood did occur.

Geological Evidence of a Worldwide Flood

Sedimentary Deposits

Sedimentary deposits cover more than seventy percent of all continental land masses. These deposits range in depth from only a few feet in some areas to thousands of feet in other areas. The depth of the basin in which the sediments were deposited determined their thickness. The strata vary in content, texture, and color in different localities, depending on the characteristics of the material from which it originally came.

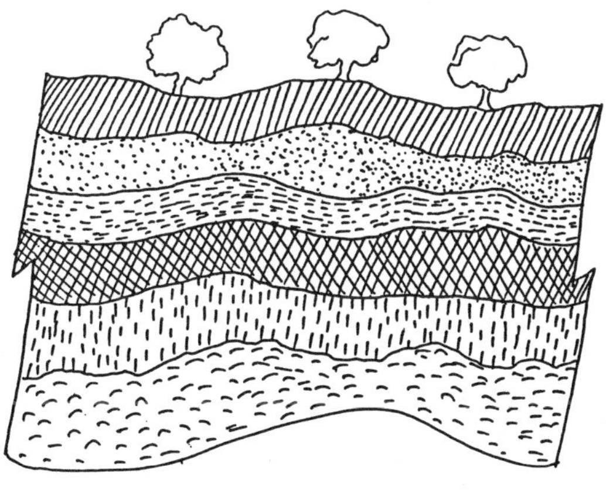

Many sedimentary deposits have solidified due to exposure to cementing agents, becoming what geologists call sedimentary rocks. Almost everyone has observed stratified layers of rock on the side of mountain passes. These are actually sedimentary deposits.

Sedimentary deposits are of great importance in studying the Flood because water formed all sedimentary deposits. Nelson describes the universality of water-laid sediments: "The earth, all around the globe, appears, wherever it is laid open, to be wholly composed of strata lying on each other in the form of so many sediments fallen down successively in water."[4]

Evolutionary scientists acknowledge that sedimentary deposits are water-laid but attribute their formation to thousands of local floods over billions of years. When a fact of science contradicts the evolutionists' concept, they immediately camouflage it with time. The central question is, however, which requires more faith—belief in God's Word, which describes a worldwide flood, or belief in billions of years of time?

One may ask how the flood could have produced the sedimentary layers. In reply, we must remember that the Noachian Flood was more than just a placid sea encompassing the globe. According to the Scriptures, part of the flood waters came from the breaking up of "all the fountains of the great deep" (Genesis 7:11). Habakkuk 3:10 associates mountain movements, which result from volcanic activity and earthquakes, with flood waters: "The mountains saw thee, and they trembled: the overflowing of the water passed by: the deep uttered his voice, and lifted up his hands on high."

Earthquakes associated with water cause tremendous geological changes, often with devastating effects on lives and property. One effect is the tsunami, or tidal wave: "Because of the origin and nature of tsunamis, they have been known to cause great damage and loss of life to certain coastal areas throughout the world."[5]

During the Flood, volcanic activity and earthquakes associated with the breaking up of "all the fountains of the great deep" must have caused tremendous surges of water, which in turn carried heavy loads of sediments with them. Scientists call them turbidity currents. As volcanic activity erupted in different localities, sediments carried by turbidity currents precipitated out and accumulated in the layers we see today.

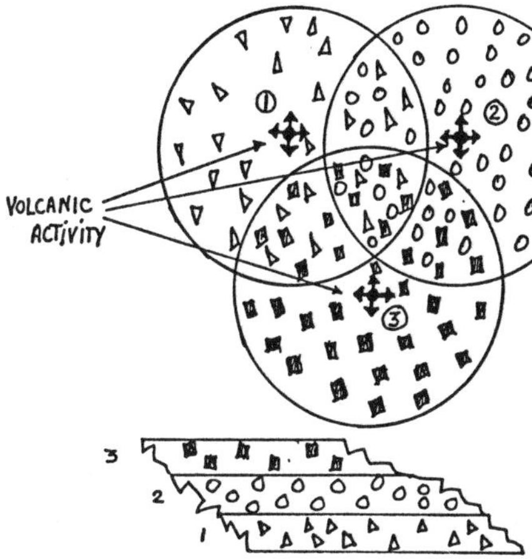

Anikouchine and Sternberg explain this sedimentary process:

> *In the turbidity flow, all sizes of sedimentary materials are transported. When suspended materials settle at the site of deposition, they are graded vertically in the order of size and density. The coarse, heavy particles settle first, the fine, light particles, last. Between flows, the pelagic interbeds deposit materials in the same scheme of formation and distribution that governs their occurrence elsewhere in the world ocean. The resulting bottom deposit exhibits an alternation in size, color, mineralogical, and biological composition.*[6]

Where did all the sediments originate? According to the uniformitarian viewpoint of evolutionists, all sediments presently upon the earth's surface came from a slow erosion of the original base rock over billions of years. According to the creationist's viewpoint the same thing occurred, except for the time scale. Instead of a slow process of erosion, the Flood had an immediate, cataclysmic erosive effect upon the earth's surface.

The Bible indicates that the surface of the antediluvian earth was very different from that of today: "For the LORD God had not caused it to rain upon the earth, and there was not a man to till the ground. But there went up a mist from the earth, and watered the whole face of the ground" (Genesis 2:5-6). In such an environment there would have been little erosion due to water. Presently, the surface of the earth (with very few exceptions) consists of a relatively thin layer of sands, silts, and clays resting on thick layers of sedimentary, metamorphic, or base rock. Since in the antediluvian world moisture rose from the lower parts of the earth to the surface, there were apparently no rock layers but only layers of unconsolidated matter between the rock base and the earth's surface.

Geological evidence of this exists in the Precambrian (oldest) sedimentary rock layer. This layer of rock, which is often very thick, is sedimentary and no fossils are found in it. This is strong evidence that before the Flood this layer was unconsolidated. These unconsolidated stratified layers without fossils would have been produced during the early formation of the earth as recorded in Genesis 1:2. Tremendous layers of unconsolidated sedimentary materials must have existed before the Flood, judging from the thick layers of sedimentary rock found throughout the earth's surface today.

When the Flood came, the extraordinary force of the surging, flowing, ebbing water along with powerful volcanic activity throughout the globe must have caused gigantic upheavals of sedimentary materials, which the waters then caught and transported. The flowing water wore and tore at the layers of unconsolidated matter for weeks and even months. Where this erosion stopped is uncertain, but it was probably at the layer in which the first fossils were found. Judging from the fossil evidence, this zone is at different depths at various places upon the earth.

Production of Coal and Oil Fields

Scientists have long advanced the production of coal and oil as evidence of earth's great age. They have more recently discovered, however, that pressure rather than age plays the primary role in the production of fossil fuels, as Melvin Cook explains:

> *Oil geologists know that the particular chemistry of oil has to do with the pressures and temperatures existing in the reservoirs or in the source rocks. Likewise, the "grade" of a coal depends on its depth of burial (Hilt's Law). Apparently all that is necessary to make coal out of wood is simply to compress it adiabatically (i.e., without loss of the heat of compression) such as to dehydrate it by the heat of compression.*
>
> *Heat is thus the factor which determines the extent of dehydration of wood, but the heat of compression is determined by the pressure.*[7]

Again, the Flood provides the most reasonable means by which enough pressure, along with little or no loss of the heat of compression, could have occurred. As previously discussed, before the Flood, a large quantity of lush, tropical vegetation grew in most areas of the globe. The Flood must have deposited billions of tons of organic matter in the many different areas where coal is today. If the flood waters rose 5,000 to 10,000 feet above sea level to cover the high mountains, then the water pressure on the earth's surface would have amounted to 300 tons per square foot, or two tons per square inch. This intense water pressure, along with the pressure of sediments deposited over the organic matter, would have been sufficient to produce coal and oil fields as we know them.

Coal balls provide a strong indication that coal was formed during the Flood. These are accumulations of plant or animal fossils cemented together. Coal balls have been found in coal beds that contain several types of marine animals such as sponges and corals.[8] This indicates that coal was formed from plant matter rapidly transported by water, in this case, sea water.

The uniformitarian explanation for coal formation is the "peat bog theory." According to this theory, great forests of vegetation grew in the same locality over millions of years. Local floods intermittently covered the organic matter, eventually producing the coal beds of today.

The biggest blow against the uniformitarian view of coal formation lies in the very definition of uniformitarianism, namely, that "the present is the key to the past." According to uniformitarian theory, there should be some location—actually many locations because there are many coal beds—where coal is presently being formed from swamps and peat bogs by present-day processes. Yet there is no known location where coal of any kind is forming today.

Production of Fossils

Upon death, most plants and animals are eaten by scavengers or break down into their chemical components and simply disappear.[9] Only those organisms or parts of organisms that are buried quickly and are protected from decay stand a chance of becoming fossils.[10] Dott and Batten explain the necessity of quick burial:

> *Another important factor essential for preservation is quick burial. Several conditions are involved here. The most important is that if the organism is left uncovered it could fall prey to such vicissitudes as scavengers, erosion, action by currents, streams, or wind. Also the or-*

ganism must be sealed tightly in the sediment to prevent decay and dissolution by ground water. It follows that there is a very limited number of optimum conditions available for rapid burial and post burial circumstances.[11]

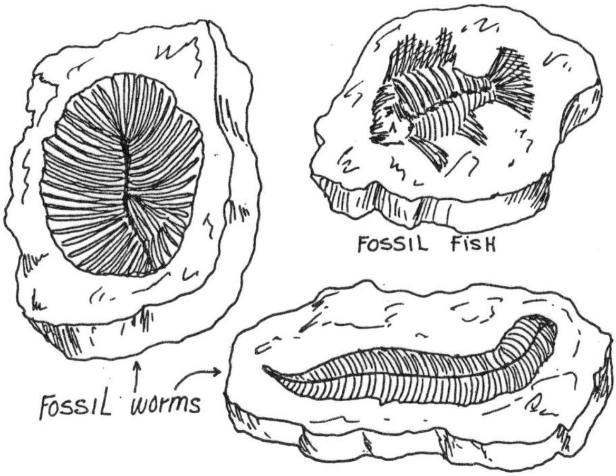

The most common method of fossilization is burial in sediments resulting from flood waters. Evolutionists Dodson and Dodson discuss the method by which an aquatic organism is transformed into a fossil, describing the key roles played by sedimentary deposition and water:

> *When aquatic organisms die, they may fall into sedimentary deposits in which the bodies are protected from scavengers and from oxidation. The soft parts of the body gradually decay and are carried away by the seepage of water. Bones and other hard parts may remain, or they may be replaced particle by particle by minerals in the water.*[12]

Every fossil speaks of death, and in most cases, cataclysmic death. Again, the Noachian Flood is in perfect accord with all of these requirements.

Polystrate Fossils

Polystrate fossils are fossils that extend through more than one sedimentary layer (strata). The uniformitarian concept of slow deposition—an estimated one foot per 2,000 years—cannot account for the existence of polystrate fossils.

As previously stated, an organism must be rapidly buried or in some other way protected from scavengers and decomposition; yet, fossilized plants and animals have been found that extended vertically across several layers of sediment. According to the dating of the layers in which the polystrate fossils were found, the layers were supposedly laid down slowly, that is, over millions of years.[13]

An interesting example of polystrate fossils is the occurrence of upright tree stumps ten feet and higher in coal seams. According to the uniformitarian, it took millions of years to produce the coal seams. If this be true, how could the tree stumps have remained upright, enduring millions of years exposed to the elements, without completely rotting away?

Marine Deposits on Mountains

Further evidence of a worldwide flood is the interesting discovery of marine fossils on most mountain ranges across the world. This discovery is not new. Even Charles Darwin in the 1800's noted the discovery of marine fossils in the Andes Mountains.[14]

Evolutionist David Merrell comments on the implications of marine deposits in high altitudes:

> *The discovery of the fossil remains of marine organisms like corals and sea urchins far inland in Indiana or 20,000 feet up in the Himalayas has far-reaching geological implications, for at one time Indiana must have been covered by the ocean, as were the Himalayas.*[15]

Though evolutionists may deny that a universal flood caused the marine deposits in the mountains, they must admit that at one time water in fact covered the mountains containing such deposits.

Fish Deposits

Numerous deposits of fossilized fish exist throughout the world in many amazing places such as the sides of mountains. Under present circumstances, it is extremely rare for a whole vertebrate fish to become entombed and fossilized. One may expect to find a few teeth, a bone or two, or even an occasional spine, but not the entire skeleton completely preserved. Yet fish from ancient times exist in marvelous states of preservation in countless deposits throughout the world. In fact, they have been discovered in whole shoals many miles in extent.

Evolutionary scientists have no explanation as to how these came to be. According to the uniformitarian concept, the present is the key to the past, but where can one find a fish deposit that is presently forming?

An underwater shock will readily kill aquatic fauna. For example, exploding dynamite creates shock waves in water that quickly kill the fish in the immediate area of the blast. Geological evidence indicates numerous volcanic explosions occurred in the flood waters, and this would have immediately destroyed large numbers of aquatic organisms.

Though evolutionists may deny the occurrence of a universal flood, all the geological evidence demands such an explanation.

What Happened to the Flood Waters?

At this point, we may well ask what happened to the flood waters that totally surrounded the globe. How did the waters abate and the dry land appear? Again the Bible supplies the answer. Genesis 8:3 states, "And the waters returned from off the earth continually: and after the end of the hundred and fifty days the waters were abated." Psalm 104:6-10 describes how the waters ran off continually:

> *"Thou coveredst it with the deep as with a garment: the waters stood above the mountains. At thy rebuke they fled; at the voice of thy thunder they hasted away. They go up by the mountains; they go down by the valleys unto the place which thou has founded for them. Thou hast set a bound that they may not pass over; that they turn not again to cover the earth. He sendeth the springs into the valleys, which run among the hills."*

These verses of Scripture indicate that after the Flood, God caused the mountain ranges and continental masses to form as we know them today. Psalm 104:8-9 in the *New American Standard Bible* states: "The mountains rose; the valleys sank down to the place which Thou didst establish for them." God then caused the waters to flow to the ocean basins and commanded them, by a covenant with Noah, "That they may not return to cover the earth."

There is certainly ample water in the oceans to have caused the Flood. If the water of the oceans were spread evenly over the entire face of the earth, it would produce a layer over a mile deep.

Psalm 104:10 describes a portion of the hydrological or water cycle (atmospheric water vapor to precipitation to springs and streams to ocean to evaporated atmospheric water). Apparently, after the Flood, God implemented the present-day hydrological cycle along with the modern seasons. (See Genesis 8:22; Ecclesiastes 1:7.)

The Rainbow

The rainbow is a good example that the Bible is in perfect harmony with the laws of science. In Noah's day, man had not progressed to the point in science to understand that the rainbow is formed by sunlight passing through atmospheric water vapor. Noah knew nothing about the prism-like effect that water has upon white light. Moses, the writer of Genesis, did not know that scientifically speaking, there could have been no rainbow before the institution of rain at the Flood. Yet, under divine inspiration, Moses recorded that God instituted the rainbow after He instituted rain.

Genesis 2:5 indicates that God had not caused it to rain on the earth before the Flood; therefore, man had never seen a rainbow. After the Flood, God started what we know today as the hydrological cycle. How could God have used the rainbow as a sign of His new covenant with man if there had been rain before the Flood? There would have been thousands of other rainbows before the Flood, making it unsuitable as the sign of God's promise to Noah that He would never again cause a worldwide flood.

Footnotes
Chapter Two

[1]Fred Wabhofsky, "Noah, The Flood, The Facts," *Reader's Digest*, Vol. III, No. 665 (Sept. 1977), p. 129.

[2]Byron C. Nelson, *The Deluge Story in Stone* (Bethany Fellowship Publishers, 1968), p. 166.

[3]Nelson, p. 168.

[4]Nelson, p. 27.

[5]William A. Anine and Richard W. Sternberg, *World Ocean, An Introduction to Oceanography* (Prentice-Hall, 1973), p. 128.

[6]*Ibid.*

[7]Melvin A. Cook, "Formation and Dating of Oil and Coal Deposits," *Bible Science Newsletter*, Vol. VIII (Jan. 1, 1970). p. 1.

[8]L. R. Moore, "Some Sediments Closely Associated with Coal Seams," *Coal and Coal-Bearing Strata*, ed. Duncan Murchison and Stanley Westoll (American Elsevier Publishing, 1968), p. 108.

[9]Peter Stoler, "Puzzling Out Man's Ascent," *Time Magazine* Vol. 110, No. 19 (Nov. 7, 1977), p. 69.

[10]*The World Book Encyclopedia*, (1960), VI, 364.

[11]Dott and Batten, p. 569.

[12]Dodson and Dodson, p. 79.

[13]*Science and Creation* (Creation Science Research Center, 1973), p. 161.

[14]Roy A. Gallent, *Charles Darwin, The Making of a Scientist* (Doubleday and Company, 1972), p. 109.

[15]Merrell, p. 40.

Chapter 3

THE FOSSIL RECORD

Fossils are the remains, impressions, or traces of animals and plants that have been preserved within the earth's crust. Fossils provide the only available record of the plants and animals of times past. Most scientists agree that the only direct evidence for or against evolution is in the fossil records. All other evidence is indirect.

As we look into the records of organic history, we find that many ancient species of animals and plants are no longer in existence today. Are those extinct organisms an indication of evolution as evolution scientists say or of a worldwide catastrophic flood as creation scientists say? To answer this question, we must look at what scientists call the geologic column.

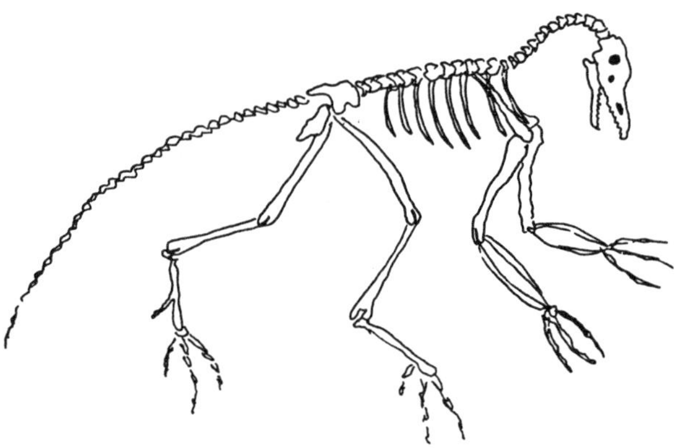

The Geologic Column

Evolutionary scientists have pieced together a vertical sequence of strata—a composite geologic column in which different local rock strata are arranged according to their relative age and then superimposed on the next older sequence.[1] This has resulted in the following geologic time scale or geologic column:

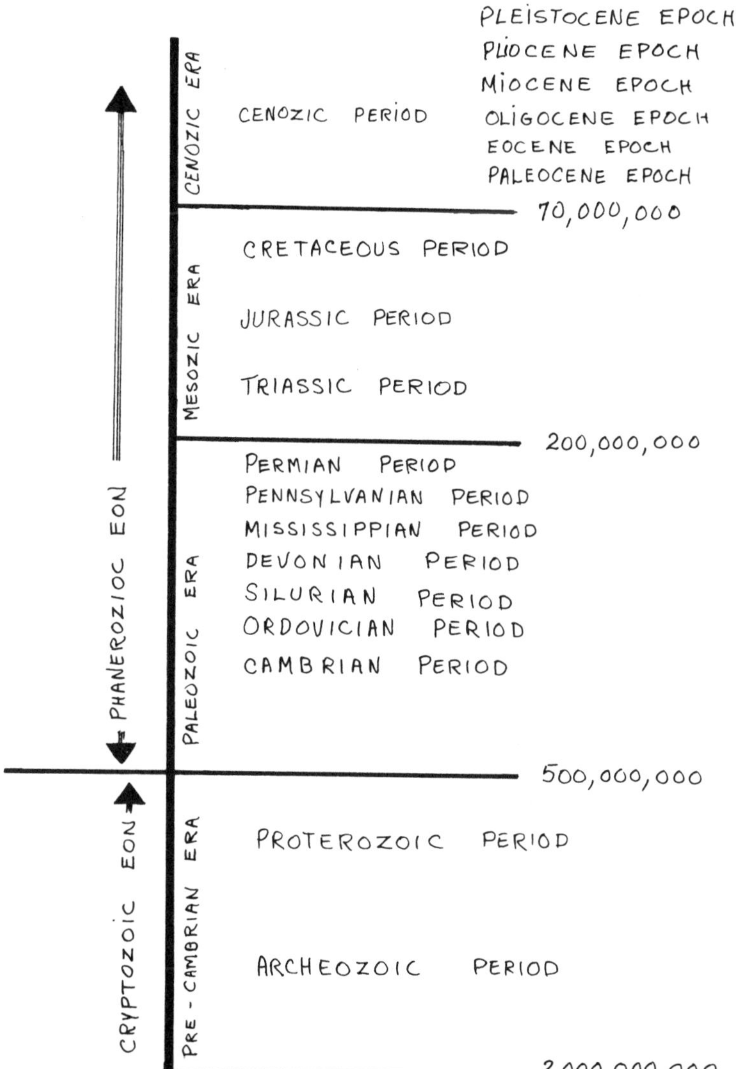
PLEISTOCENE EPOCH
PLIOCENE EPOCH
MIOCENE EPOCH
OLIGOCENE EPOCH
EOCENE EPOCH
PALEOCENE EPOCH
CENOZIC ERA
CENOZIC PERIOD
70,000,000
MESOZIC ERA
CRETACEOUS PERIOD
JURASSIC PERIOD
TRIASSIC PERIOD
200,000,000
PHANEROZIOC EON
PALEOZOIC ERA
PERMIAN PERIOD
PENNSYLVANIAN PERIOD
MISSISSIPPIAN PERIOD
DEVONIAN PERIOD
SILURIAN PERIOD
ORDOVICIAN PERIOD
CAMBRIAN PERIOD
500,000,000
CRYPTOZOIC EON
PRE-CAMBRIAN ERA
PROTEROZOIC PERIOD
ARCHEOZOIC PERIOD
2,000,000,000

Theoretically, the geologic column yields a succession of organisms beginning with the simple at the bottom and progressing successively upward to the complex. This so-called "fact" is often presented as definite proof of organic evolution; yet there is no place on the earth's surface where such a geologic column exists. Evolutionist Merrell acknowledges as much:

> *If it were possible to find a place where deposition of sediments had been continuous since the formation of the earth in its present structure, the strata would form a complete geological column, and the included fossils would furnish a fairly good record of the forms of life that had existed during this period. Although some deposits are thousands of feet thick, no such complete geological column is known.*[2]

According to the uniformitarian theory, slow deposition took place over a period of more than three billion years. If this be true, a complete geologic column would be approximately 100 miles thick.

If a complete geologic column does not exist anywhere on earth, how did evolutionists come up with the composite geologic column found in all textbooks dealing with evolution? Evolutionist Hotton explains: "Lowland deposits are smaller and less continous than marine beds, and so it is often necessary to supplement gaps in the record of one continent with information from another."[3] Gaps in the record of one continent are supplemented with information from other continents. In other words, evolutionists find fossils from many different locations and arrange them, along with the strata in which they are found, in a hypothetical sequence of increasing complexity. They in turn use this geologic column as evidence of evolution. This is clearly a form of circular reasoning. Evolutionists Dunbar and Waage provide a vivid example of this circular argument:

> *We now know that different kinds of animals and*

> *plants succeeded one another in time because life has continuously evolved: and inasmuch as organic evolution is worldwide in its operation, only rocks formed during the same age could bear identical faunas.*[4]

Evolutionary dating methods provide another example of circular reasoning. Geologists date the geologic columns by the use of index fossils, while biologists determine the age of the index fossils by the strata in which they are found. R. H. Rastall admits the circular nature of this process:

> *It cannot be denied that from a strictly philosophical standpoint geologists are here arguing in a circle. The succession of organisms has been determined by a study of their remains embedded in the rocks, and the relative ages of the rocks are determined by the remains of the organisms that they contain.*[5]

To illustrate how circular reasoning works, let us suppose that we find a fossilized bone and take it to a well-known paleobiologist to determine its age. After examining the fossil he declares it to be approximately 200 million years old. We then ask him how he arrived at this figure. Did he use the potassium-argon dating method or the rubidium-strontium dating method?

Surprisingly, the paleobiologist answers that he did not use a dating method. He explains that this fossil is always found in the same series of rock strata of the geologic column and that historical geologists have determined that particular series to be 200 million years old.

We then rush to the historical geologist. He examines the fossil and says it is from the Middle Triassic Series of the Triassic Period of the Mesozoic Era and so is 200 million years old. We then ask how he knows the age of that particular series of rock strata. Does he use the uranium-lead dating method or the lead-alpha dating method? He explains that paleobiologists have found this type of fossil to be 200 million years old; therefore the strata in which the fossil was found must also be 200 million years old.

Both scientists have assumed each other's conclusions. They have only assumed, not proven, the theory of evolution.

At this point some may argue that geologists actually can use scientific dating techniques to ascertain the ages found in the geologic column. For more information about the available dating techniques and their reliability, see Chapter 10.

Patterns in the Geologic Column

There is seemingly a certain sequence or trend in the placement of various kinds of organisms in the geologic column. Evolutionists often present this evidence in support of their theory. There are, of course, countless exceptions to the postulated sequence, yet many fossil organisms do tend to follow a pattern from the top of the geologic column downward. The basic pattern is as follows:

Man, Elephants, Horses, Apes, Birds
Dinosaurs and Reptiles
Amphibians
Fish and Sharks
Corals, Trilobites, Molluscs

From a quick glance, the pattern in the geologic column seems to give credence to the evolution theory; yet, as we will discuss later, no series of fossils throughout the entire fossil record manifests a transition from one kind of organism into another kind, which would be required to prove evolution conclusively from the fossil record.

Furthermore, we would expect this very sequence, if we believe in creation, followed by a worldwide cataclysmic flood. For example, at the very bottom of the sequence are the corals, trilobites, and molluscs. These organisms lived in the mud or near the mud at the bottom of oceans and lakes. When the Great Flood began, sediments would have covered these organisms first. Next to be covered would have been those organisms which inhabited the water, such as various different kinds of fish and sharks. As the Flood progressed, the rising water and accompanying turbidity currents would have next entombed the amphibians—those animals which live in a transition zone, partially in the water and partially on land. The rising water would have next entombed the slower-moving land animals, such as dinosaurs and reptiles. Finally the flood waters and accompanying sediments would have trapped the quicker-thinking and faster-moving land creatures such as horses, elephants, apes, birds, and man.

The pattern found in the geologic column is no evidence of evolution because there is no record of transition from one kind of organism into another. Rather, the record indicates the natural deposition of living things as they were entombed by a worldwide flood.

The Sudden Appearance of Complex Organisms

For evolutionists, one of the greatest dilemmas of the fossil record is the sudden appearance of a great variety of highly complex organisms in the Cambrian period (the oldest rock system containing fossils). According to the creation concept, we would expect the lowest layers of sedimentary rock (pre-Cambrian) to be void of fossils (Genesis 1:1-3). Going up the geologic column, we should find a sudden outburst of fossils due to the Noachian Flood. This, of course, is exactly what we do find in the fossil record.

Evolutionist Dobzhansky acknowledges that the Cambrian contains representatives of most major groups of animals now living in modern seas. Furthermore, some of the Cambrian creatures were quite elaborate and advanced in body structure.[6] Hotton also indicates as much: "The diversity and relatively sudden appearance of the Cambrian fauna have been interpreted to mean that hard parts were acquired independently by a large variety of organisms in a relatively short time."[7]

The Cambrian actually contains these phyla:

Protozoa—Single-celled organisms
Porifera—Sponges
Coelenterata—Jellyfish, anemones, and corals
Archeocyatha—Sponge or coral-like creatures
Brachiopoda—Bivalves superficially resembling clams (lampshells)
Mollusca—Clams, snails, and octopuses
Arthropoda—Jointed-leg creatures such as insects, crustaceans, and trilobites
Echinodermata—Spiny-skinned animals such as starfish, sea urchins, and sea lilies
Chaetognatha—Arroworms
Annelida—Segmented worms

Leclercq has even demonstrated the evidence of vascular plants in the Cambrian.[8] Clifford Burdick, a consulting geologist, has reported the discovery of pollen from pines and other plants in all sedimentary layers exposed in the Grand Canyon, from the recent Cenozoic all the way down to the earliest Cambrian and even pre-Cambrian rocks.[9]

According to evolution, however, there should be a gradual progression from the least complex to the highly complex. Consequently, evolutionist Dickerson is forced to state that the first billion pages in the book of earth's history are almost completely missing.[10]

How do evolutionists explain the sudden appearance of complex organisms in the Cambrian? They often say that organisms existing before the Cambrian Period had no preservable skeletons because little or no calcium was present in the sea water. This explanation is merely an assumption, and one that contradicts uniformitarianism. How can they prove there was no calcium in sea water at that time? Furthermore, calcium was not the only material available to the Cambrian organism for support, strength, and hardness. Their exo-skeletons were also composed of chitinous, siliceous, and calcareous hard parts. If evolution did occur, then, there should be at least some fossil records of a transition from very simple organisms to the complex organisms actually found in the earliest Cambrian.

The father of evolutionist theory, Charles Darwin, aptly summarizes evolution's dilemma with respect to the sudden appearance of complex organisms in the fossil record:

> *The abrupt manner in which whole groups of species suddenly appear in certain formations, has been urged by several palaentologists, for instance, by Agassiz, Pictet, and by none more forcibly than by Professor Sedgwick, as a fatal objection to the belief in the transmutation of species. If numerous species, belonging to the same genera or families, have really started into life all at once, the fact would be fatal to the theory of descent with slow modification through natural selection.*[11]

Gaps in the Fossil Record

The innumerable gaps found throughout the entire fossil record provide one of the strongest objections to the theory of evolution. Almost everyone has heard of the proverbial "missing link"; there are, in fact, millions of missing links. Even Darwin was puzzled by this enigma: "First, why, if species have descended from other species by fine graduations, do we not everywhere see innumerable, transitional forms? Why is not all nature in confusion, instead of the species being, as we see them, well defined?"[12]

As Darwin recognized, one of the greatest deficiencies in his theory is that different kinds (species) of animals and plants are well defined with no transition from one to the other. For example, there are many different variations of cats and many variations of dogs; yet there is not one transitional form between cats and dogs.

The same holds true in the fossil record. There are innumerable fossils of various kinds of organisms; yet there are no transitional forms that link two different "kinds" together. This is exactly what God established in Genesis 1:24-25: "And God said, Let the earth bring forth the living creature after his kind, cattle, and creeping thing, and beast of the earth after his kind: and it was so. And God made the beast of the earth after his kind, and cattle after their kind, and every thing that creepeth upon the earth after his kind: and God saw that it was good."

Most evolutionists recognize the problem that the lack of transitional forms presents to the theory of evolution. For example, evolutionist Merrell states:

> *The origin of the higher taxonomic categories has presented a problem because new orders and classes generally appear suddenly in the fossil record, without evidence of intermediate fossil types. If evolution is a gradual process as both Darwin and modern theory hold, then it might be expected that fossils connecting*

> *different orders would be found as evidence of the gradual evolutionary transition from one group to another.*[13]

During the formulative years of evolutionary theory, many scientists, including Darwin, felt that they would eventually find transition forms between different kinds of organisms. Time has proven fatal to this idea, however. Although many new fossils have been found, the gaps in the fossil record are no narrower today than when Darwin and other evolution scientists first made their statements about them.

Evolutionists often sound as if they are looking for only one missing link in the entire phylogenetic sequence. In reality, they are looking for thousands of missing links in the chain that would relate every kind of organism to other kinds of organisms. These so-called "missing links" would appear in the following manner:

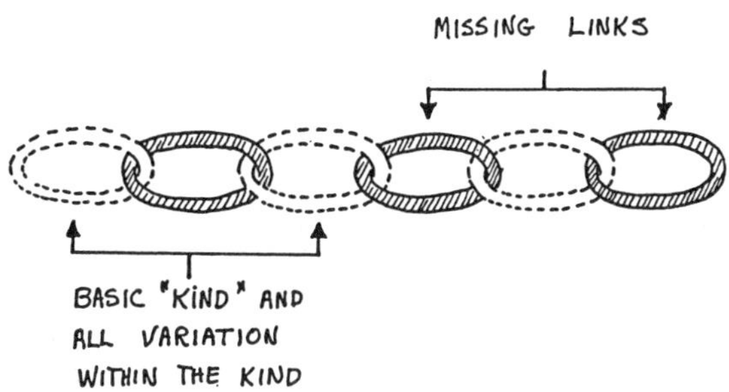

If evolution did occur in the past, the fossil record should indicate numerous types of monstrosities and terribly-formed organisms. These forms would have eventually lost out in their struggle for survival as better-adapted forms were selected, yet their fossil remains should still be present. What does the fossil record actually show? It reveals completely-formed animals and plants in every geologic column, even in the Cambrian. Every fossil indicates that it was perfectly designed and adapted for its particular environment.

If natural selection produced the modern species of today, then where are the fossils of the transitional forms that became extinct? Modern plants and animals, as well as all fossils from the Cambrian to the Cenozoic, always exhibit perfect design rather than haphazard formation.

One of the strongest evidences of evolution past and present should be monstrosities in the fossil record and in nature today. We simply do not find them, however. On the other hand, the evidence of design in living systems, as depicted by completely-formed animals and plants, is a strong indication of creation.

Evolutionists often describe how the appendage of one organism evolved into the appendage of another, yet the fossil record is void of such transitions. When someone challenges the evolutionist to explain, he often replies, "Conditions were not right for the fossilization of transition forms." While this could be true in some situations, there should be at least a few transition forms in the fossil record. If we can find the original form and the supposedly evolved form, then why do we never find the intermediate forms?

For example, the fins of certain fish supposedly evolved into the feet and legs of the first amphibians, yet there is no fossil evidence to support this theory. In the fossil record are numerous examples of fish with fins and numerous examples of amphibians with feet, but not one single transition form between the two.

The ideal conditions for fossil formation would have been present for these supposed intermediates. The intermediates, just as the fish and amphibians, would have spent considerable time at the edges of lakes, streams, and ponds. Evolution would require innumerable years for the transition of fins into feet and legs, and uniformitarianism would require innumerable local floods in that time. These floods would certainly have captured, buried, and fossilized many of the transition forms.

The following are quotations from many leading evolutionists concerning the gaps in the fossil record. We should remember that evolutionists cite the fossil record as the strongest evidence for their theory.

> *Despite the fossil record's being incomplete and in many ways baffling, we may entertain a working hypothesis that the evolutionary events for which this record stands were brought about by natural selection in response to environmental opportunity.*[14]
>
> *First of all, it must be reemphasized that there is not a continuum in the pattern of variation. There are, for example, no individuals who are intermediate in their traits between a house cat and a dog. Even in cases where the resemblance is much closer than that between a dog and a cat, intermediates do not exist.*[15]
>
> *For recovery, a fossil deposit must remain well buried just the right length of time, and then must again be near the surface and partly exposed by erosion within a short time before a bone digger happens to come along. Small wonder that there are deficiencies in the fossil record. When you consider this series of chances it is wonderful that the record is not all gaps.*[16]
>
> *Thus far, we have seen that the fossil record is biased and filled with gaps that produce, at best, only a fuzzy picture of life in the past.*[17]
>
> *Sometimes the study of the fossil remains of extinct organisms provides clues to the evolutionary history of a group of species, but the fossil record is always incomplete and often altogether lacking.*[18]
>
> *In fact, so sparse is the fossil record, despite the tens of thousands of fossils that are stored and studied in museums, that to get a complete picture of evolution is a little like trying to reconstruct* Gone With The Wind *from the scraps on the cutting-room floor.*[19]
>
> *The clues to man's past are chiefly fossils, a farrago*

> *of frequently undecipherable—and occasionally contradictory—bits of evidence that often raise more questions than they answer.*[20]
>
> *We have reviewed the fossil record, and seen that it is the most impressive evidence of evolution, in spite of its grave deficiencies.*[21]
>
> *Many gaps exist in the record, especially before animals developed hard parts.*[22]

As the preceding quotes demonstrate, the so-called evidence for evolution in the fossil record has in reality become evidence against the theory.

Plant Evolution

Over 100 years ago, Charles Darwin called the origin of flowering plants (angiosperms) an "abominable mystery."[23] It would seem that over one hundred years of searching the fossil record would have solved this mystery, yet Dodson and Dodson stated in 1976: "How the vascular plants arose from their algal ancestors is not known, and the fossil record throws little light upon this important question."[24] In fact, Merrell calls the origin of the Angiosperm, which is dominant in the present day flora, a "complete mystery."[25] Obviously evolutionists today are no closer to proving the evolution of flowering plants than Darwin was in 1859.

As mentioned earlier, pine pollen, along with other plant pollen, has been found in every sedimentary layer exposed in the Grand Canyon.[26] These layers of rock range from pre-Cambrian to recent Cenozoic. How did pine pollen get into the Cambrian and pre-Cambrian rock strata if the pine tree did not evolve until much later? Of course, we would expect such pollen dispersion as a result of a worldwide flood.

Archaeopteryx

Evolutionists assume that birds evolved from reptilian ancestors. They usually present *Archaeopteryx*, an extinct flying animal, as the "missing link" that supports this assumption.

Archaeopteryx had true feathers, a wishbone,[27] and wings, which are all characteristics of birds. *Archaeopteryx* also had teeth, thus giving it a characteristic of reptiles. By itself, the possession of teeth cannot prove that *Archaeopteryx* evolved from reptiles, because the fossil record shows that many reptiles did not have teeth. The same is true for amphibians, fishes, birds, and mammals. Some have teeth and others do not.

Another so-called reptilian characteristic is the claws in the wings of *Archaeopteryx*, yet the hoatzin, a modern bird of South America, also possesses claws on its wings when young. These claws enable it to climb on tree branches. If the hoatzin were extinct today and one of its fossils appeared in the appropriate strata, evolutionists would call it an intermediate between reptiles and birds. Because it is alive today, they recognize it as completely bird.

Another assumption concerning *Archaeopteryx* is that its feathers evolved from the scales of reptilian ancestors, but feathers come from a totally different layer of skin than scales do. In fact, the feathers of *Archaeopteryx* are entirely comparable to those of modern birds.[28]

Evolutionist Lewontin describes the changes necessary for a bird to evolve from a reptile:

> *In the course of the evolution of birds from reptiles there was a successive alteration of the bones, the muscles and the skin of the forelimb to give rise to a wing: an increase in the size of the breastbone to provide an anchor for the wing muscles; a general restructuring of bones to make them very light but strong, and the development of feathers to provide both aerodynamic elements and lightweight insulation.*[29]

What are the chances that a reptile would experience a series of mutations giving it all the needed requirements to fly? When we undertake this analysis, we quickly discover the statistical impossibility of a bird evolving from a reptile.

If mutations produced everything needed for flight except one thing, then natural selection would immediately eliminate that animal. Although a junk yard may have all the necessary parts of a car, until these parts are assembled and made to work in unison they are merely junk. Likewise, natural selection would have elimated anything less than a complete bird. As a misfit in the biological community, it could not have survived to evolve further.

In conclusion, *Archaeopteryx* is simply an extinct order of birds.

The Horse

Of all fossil pedigrees, evolutionary literature most often mentions the horse.[30] Evolutionists contend that they have proven their theory by tracing the phylogeny (evolutionary history) of the horse through the entire Cenozoic Period. It is supposedly one of the most thoroughly known phylogenies of any vertebrate.[31] As Merrell states, "Horses have left behind the most complete sequence of fossils yet discovered. Their history has therefore been worked out in greater detail than that of any other group."[32]

According to evolutionists, the fossil record is nearly complete in regard to the evolution of the horse[33] and allegedly demonstrates a transition from a five-toed, rodent-like ancestor to the present-day 2,000 pound horse.[34] If this is the case, then evolutionists have a reasonably strong argument regarding change within a family; yet, even so, this argument would not prove the theory of evolution conclusively because it does not demonstrate change from one phylum to another. On the other hand, there are strong indications that the supposed evidence of the evolution of the horse is simply a vain attempt by the evolutionist to prop his ailing theory.

In fact, the men who first worked out the broad outlines of the horse family history were interested above all in testing Darwin's theory of evolution.[35] When Darwin's theory was first gaining popularity in scientific ranks, it direly needed supporting evidence. That evidence was provided in a hastily compiled phylogenetic tree demonstrating the alleged history of the horse as George Gaylord Simpson says:

> *Within twenty years of the publication of* The Origin of Species *the paleontologists were able to demonstrate by a nearly continuous series of intermediate forms that the little animal Hyacotherium, found by Colchester and Richardson a generation earlier, which was an English representative of the eohippus group, was the ancestor of our living horse.*[36]

Almost every evolutionist begins the family tree with *Hyracotherium* (also called *Eohippus* or "dawn horse"), yet *Hyracotherium* had no resemblance to the modern horse.[37] Its teeth were small, low, and cusped and really more like monkey teeth than horse teeth, which are great, ridged, grinding prisms. The little head of *Hyracotherium* had large eyes set midway from its snout to its ears rather than far back in the skull as are the relatively small eyes of the modern horse.

Furthermore, *Hyracotherium* was considerably more like

a condylarth than a horse. Its back was arched and flexible compared to the perfectly straight, rigid backbone of the modern horse. In fact, Simpson says that *Hyracotherium* had high hindquarters which gave it a striking un-horse-like, almost rabbit-like, appearance. *Hyracotherium* had dog-like pads on its feet rather than hooves.[38] In short, "the earliest Eocene equines were so unlike the modern horses that they were called *Hyracotherium* because of their rodentlike appearance."[39]

The contrast could go on and on. As Simpson summarizes: "The large eye is still set in the middle of the skull (from front to back), the orbit is still open behind, and the snout is still no longer than the brain case. These proportions are so different from *Equus* that the head of Eohippus, when correctly restored, does not look like a small horse's head."[40]

With such vast difference between the horse and *Hyracotherium*, why do evolutionists place it at the root of the horse group? The answer is apparently tradition or custom:

> *We now know that the evolution of the horse did not always take a simple path. In the first place it is not clear the Hyracotherium was the ancestral horse. Thus Simpson (1945) states, "Matthew has shown and insisted that Hyracotherium (including Eohippus) is so primitive that it is not much more definitely equid that tapirid, rhinocerotid, etc., but it is customary to place it at the root of the equid group."*[41]

In other words, the evolutionist had to start somewhere; he therefore rummaged through the fossils and did the best he could to create a family tree for the horse.

Although evolutionary literature does not publicize it, most evolutionists no longer advocate the traditional single-line lineage of the horse (*Hyracotherium* to *Orohippus* to *Mesohippus* to *Merychippus* to *Plionhippus* to *Equus*). On the contrary, they tend to believe that the horse split up into various groups, which evolved separately and in somewhat different ways.[42] Simpson notes that the popular literature on the subject is usually inaccurate:

> *It is sad that this introduces possible confusion into the story, but there is not much point in criticizing nature for something that happened some millions of years ago. It would also be foolish to try to ignore the complications, which did occur and which are a very important part of the record. Most accounts of horse evolution have simplified the situation to the point of definitely falsifying it.*[43]

The drama of horse evolution has become exceedingly complex as more material has been collected and incorporated. The so-called branches of the family tree have increased in size and complexity until the shape is presently more like a bush than a tree.[44] Evolutionists have classified over twenty genera and a much larger number of species of horses,[45] but evolutionists do not include them as links because they do not fit the preconceived idea of how the horse supposedly evolved.

In addition, the fossils of the traditional line of descent for the horse have not all been found in one area of the world, which would be necessary to support evolutionary theory, but rather have been found from China to western Europe to the United States.[46] To further complicate the situation, three-toed horses have been found side by side with one-toed horses. It is rather hard for a three-toed horse to evolve into a one-toed horse when both are grazing side by side. Evolutionists often state that the splint bones of modern horses are vestiges of the four toes of the ancestors, yet *Hyracotherium*, the so-called earliest horse, had splint bones.[47]

In conclusion, the best evidence of horse evolution is the pictorial display in biology books, and even this is simply a projection of the evolutionists' imagination. Once again, evolutionary "fact" has become fiction.

Footnotes
Chapter Three

[1]Dunbar and Waage, p. 10.

[2]Merrell, pp. 40-41.

[3]Hotton, p. 53.

[4]Dunbar and Waage, pp. 10-11.

[5]R. H. Rastall, "Geology," *Encyclopedia Brittanica* (1956), X, p. 168.

[6]Dobzhansky, p. 288.

[7]Hotton, p. 48.

[8]J. Kerby Anderson and Harold C. Coffin, *Fossils In Focus* (Probe Ministries International, 1977), p. 71.

[9]*Science and Creation* (Creation Science Research Center, 1973), p. 160.

[10]Richard E. Dickerson, "Chemical Evolution and the Origin of Life," *Scientific American*, Vol. 239, No. 3 (Sept. 1978), p. 70.

[11]Charles Darwin, *The Origin of Species by Means of Natural Selection* (Crown Publishers,), p. 309.

[12]Horatio Hackett Newman, *Evolution, Genetics, and Eugenics*, 3rd ed. (University of Chicago Press, 1932), p. 590.

[13]Merrell, p. 295.

[14]Dobzhansky, p. 366.

[15]Merrell, p. 80.

[16]George Gaylord Simpson, *Horses* (Oxford University Press, 1951), p. 95.

[17]Dott and Batten, p. 600.

[18]Francisco J. Ayala, *The Mechanisms of Evolution*, p. 68.

[19]Hotton, p. 42.

[20]Peter Stoler, "Puzzling Out Man's Ascent," *Time Magazine*, Vol. 110, No. 19 (Nov. 7, 1977), p. 69.

[21]Dodson and Dodson, p. 421.

[22]*The World Book Encyclopedia* (1960), VI, p. 364.

[23]Dodson and Dodson, p. 150.

[24]*Ibid.*, p. 139.

[25]Merrell, p. 152.

[26]Clifford Burdick, "Microflora of the Grand Canyon," *Annual* (Creation Research Society, Annual 3, No. 1, 1966), pp. 38-50.

[27]John H. Ostrum, "A New Look at Dinosaurs," *National Geographic*, Vol. 154, No. 2 (Aug. 1978), p. 168.

[28]Dodson and Dodson, p. 218.

[29]Richard C. Lewontin, "Adaptation," *Scientific American*, Vol. 239, No. 3 (Sept. 1978), p. 213.

[30]Newman, p. 133.

[31]Dodson and Dodson, p. 86.

[32]Merrell, p. 46.

[33]*The World Book Encyclopedia* (1960), VI, p. 367.

[34]Newman, p. 136.

[35]Simpson, p. 165

[36]*Ibid.*, p. 87.

[37]*Ibid.*, p. 86.

[38]*Ibid.*, p. 117.

[39]Merrell, p. 48.

[40]Simpson, p. 118.

[41]G. A. Kerkut, *Implications of Evolution* (Pergamon Press, 1960, repr. 1973), p. 149.

[42]Simpson, p. 127.

[43]*Ibid.*

[44]Kerkut, p. 149.

[45]Dodson and Dodson, p. 86.

[46]Simpson, p. 88.

[47]James A. Michener, "Life and Death of the Dinosaur," *Reader's Digest*, Vol. 105, No. 631 (Nov. 1974), p. 87.

Chapter 4
HUMAN EVOLUTION

If evolution did occur, the most obvious evidence would be the evolutionary development of man as portrayed in the fossil record, for man is supposedly the epitome of evolutionary engineering. If man evolved in the recent past, a wealth of fossils should document his precarious journey from an apelike creature to modern man. There should be a continuous series of transitional forms from lower mammals to modern man, but to the embarrassment of evolutionists, the fossil record is void of such a portrayal of man's development.

A *Time Magazine* article has described this paucity of fossil evidence for man's evolution:

> *Still, doubts about the sequence of man's emergence remain. Scientists concede that even their most cherished theories are based on embarrassingly few fossil fragments, and that huge gaps exist in the fossil record. Anthropologists, ruefully says Alan Mann of the Univer-*

sity of Pennsylvania, are like the blind man looking at the elephant, each sampling only a small part of the total reality. His colleagues agree that the picture of man's origins is far from complete.[1]

Evolutionist Merrell reinforces this by stating that the fossil record of man is not complete enough to show exactly what has been the course of evolution leading to "Homo Sapiens."[2]

Evolutionist Dobzhansky describes the fossil record of primates as "unfortunately meager":

> *Furthermore, whatever remains are found are usually fragments of bones and teeth. It is a difficult matter to get from such fragments an idea of what the whole animal might have been like. Because of this difficulty, the disagreements between different authorities on fossil human and prehuman beings are often extreme and rather exasperating to non-specialists.*[3]

From the fragmentary fossils uncovered, evolutionists have developed many different views of evolution. Even evolutionists disagree among themselves and express doubt as to the reliability of these views. As evolutionist Sherwood L. Washburn comments, "All views of human evolution are built on seeming facts that vary widely in their degree of reliability."[4]

Despite the meager evidence and fragmentary fossil finds, evolutionists still offer their conception of prehistoric man complete with hair, eyes, ears, nose, lips, and even facial expressions. This reconstruction, of course, is a figment of their imagination and is far removed from true science, for the fossil records do not preserve these features. The numerous sketches and statues of prehistoric men found in evolutionary texts and museums border on the absurd.

Again we should note the circular reasoning employed. Evolutionists assume that evolution has taken place. In accordance with that assumption, they reconstruct fossil fragments to manifest primitive characteristics and expressions. They then display these reconstructions as evidence for the theory of evolution.

The very word *prehistoric* indicates supposition. The last 5,000 years are called the "historic" period of man. Records, documents, writings, cities, and other artifacts verify it. The word *prehistoric* means exactly what it says. It is pre-history or, in other words, the time before any historical documentation. The concept of prehistoric men is a pure assumption formulated to provide evidence to support another assumption, namely, evolution.

Amazingly, much of the evolutionary literature contains numerous "facts" about the so-called ancestors of man that no one can prove or even document. For example, an article in *Science Digest* describes how primates began to live on the ground:

> *With the trees disappearing, these animals were faced with the problem of surviving on the ground. At*

first it was merely a matter of getting from one patch of forest to another. Gradually the need for food forced them to the ground, and balanced somewhat precariously on two feet, they moved awkwardly between trees, often scrambling after rodents for dinner.[5]

How could the author of this article know with such certainty the details of the prehistoric lifestyle?

Although numerous fossil apes and fossil men have been found, only a handful of highly questionable bone fragments could possibly indicate intermediates between an ape-like ancestor and man. This scarcity seems rather strange considering the great number of years that evolutionists have been searching for intermediates. This lack of evidence of intermediates provides a strong objection to the theory of human evolution.

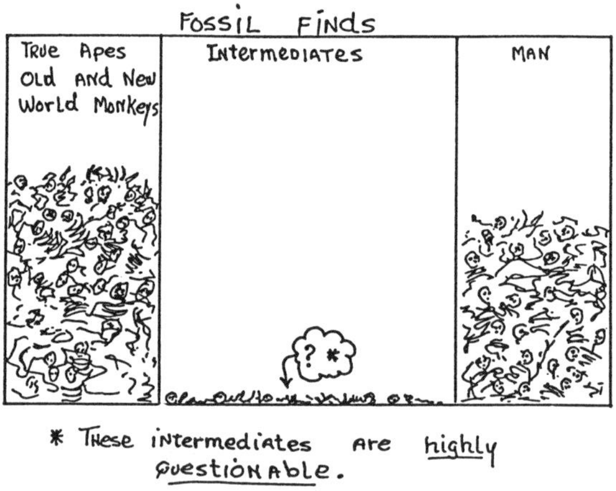

According to evolutionary theory, the major change that led to man was the shift from four-footed locomotion to bipedal (two-footed) locomotion. This shift required changes in the bones and muscles of the pelvis, legs, and feet, including realignment of the big toe and change in the angle of attachment of the skull to the spine.[6] Unfortunately for the evolutionists, not one of these changes can be documented in the fossil record.

At this point, let us look at the so-called "ancestors" of modern man.

Piltdown Man

Eoanthropus dawsoni

In 1912, Charles Dawson, an amateur collector, announced the discovery of a very ancient man from the Thames gravels at Piltdown, Sussex, England. The material consisted of portions of the cranial walls, nasal bones, a canine tooth, and part of the lower jaw.[7] When the material was reconstructed, it took the form of a brain case much like that of modern man, although thicker, with a lower jaw resembling that of a large ape. This find fulfilled evolutionists' conception of a "missing link" between man and ape.[8]

From 1912 to 1955, scientists offered Piltdown man as definite proof of man's evolution. In 1955, however, modern techniques of analysis showed that Piltdown man was unquestionably a fraud. The remains were identified as a human skull and the carefully doctored jaw and teeth of an orangutan.[9] Tests conducted on the bones showed that someone had stained them to simulate a fossil of great age. The teeth showed unnatural wear, and microscopic examination revealed file marks.[10] Needless to say, after 1955, museums and textbook authors quickly and quietly eliminated busts and drawings of Piltdown man. In this case, "fact" not only became fiction but "fake."

Nebraska Man

(Hesperopithecus haroldcookii)

On March 14, 1922, Henry Osborn, director of the American Museum of Natural History in New York City, received a package from Harold Cook, a consulting geologist.

The package contained a fossilized tooth from Nebraska. Eight days later, Osborn proclaimed that the tooth represented a new genus of anthropoid ape and proceeded to name it *Hesperopithecus haroldcookii*.[11]

In 1926, H. H. Wilder wrote a book concerning Nebraska man entitled *The Pedigree of the Human Race*.[12] At this point, it seemed evolutionists had certainly found a "missing link." Unfortunately, in 1927 more fossil material of Nebraska man was found. He turned out to be a type of extinct pig and, once again, "fact" became fiction.

Java Man

(Pithecanthropus erectus)

(Homo erectus)

In 1891, Dr. Eugene Dubois found a skullcap (calvarium), a left thigh bone, and two upper molar teeth while excavating Pleistocene deposits on the Solo River in Central Java.[13] Dubois called this material *Pithecanthropus erectus* (erect ape-man), which later became known as Java Man. Dubois estimated the cranial capacity to be about two-thirds that of modern man, although it was impossible to get a proper estimate from the skullcap alone.

Boule, one of the world's greatest authorities on fossil skulls, after thoroughly analyzing the skullcap, said it was very similar to those of chimpanzees and gibbons.[14] Ashley Montagu in his book *Man—His First Million Years* said the skullcap looked very primitive.[15]

Not all evolutionary scientists concurred with Boule and Montagu, but the real deathblow came to Dubois' Java man when he revealed in 1922 that he had found two human skulls at approximately the same level of strata. These skulls, which later became known as the "Wadjak Skulls," had a cranial capacity above that of modern man's.

It is impossible for a "missing link" to evolve into modern man when they existed side by side. Apparently, Java man was merely a large gibbon, and again evolutionary "fact" has become fiction.

Peking Man

(Sinanthropus pekinensis)

Between 1920 and 1930, the fragments of approximately 30 skulls, 11 lower jaws, and about 147 teeth were found 25 miles from Peking, China. Davidson Black christened these remains *Sinanthropus pekinensis*, later known as Peking Man. These fossil fragments became the subject of much controversy —until all of the material except two teeth mysteriously disappeared.[16]

Australopithecines

(Australopithecus africanus)
(Australopithecus robustus)
(Homo habilus)

We will not say much regarding these extinct creatures for

most evolutionists no longer regard them as ancestors of the human race. According to Dobzhansky, scientists have reached no agreement concerning the position of the *Australopithecines* among the primates.[17] According to Richard Leakey, they represent a sterile side branch of hominid evolution.[18]

The *Australopithecines* were small creatures weighing approximately 50 pounds and measuring possibly four feet tall when standing.[19] Their brain capacity was only 400 to 600 centimeters, less than one half that of modern man. According to Dodson and Dodson, the skull of *Australopithecus* resembled that of a modern chimpanzee.[20]

Zinjanthropus
(Zinjanthropus boisei)

In 1959, L.S.B. Leakey and his wife found 400 fragments of a skull in the Olduvai Gorge of Kenya, Africa. They reconstructed the skull and placed it in the line of man's descent. They proclaimed it to be—at two million years of age—man's oldest ancestor. Magazines, television, radio, and even high school biology textbooks set forth this so-called fact concerning *Zinjanthropus*.

In 1964, however, many anthropologists—including Leakey himself—changed their minds and decided that *Zinjanthropus* was nothing more than an ape, probably an *Australopithecine.*[21] By 1970, three different artists had depicted the 400 fragments of *Zinjanthropus boisei* in three very different ways for scientific articles.

Neanderthal Man

(Homo neanderthalensis)

A skullcap, together with a few ribs and limb bones, were excavated in 1856 from a cave in the Neanderthal Valley of Germany. These remains became known as Neanderthal Man and were popularly regarded as the "prehistoric man."[22]

Upon reconstruction of the skeleton, scientists found it had curvature of the spine and also curiously curved thigh bones.[23] Following this discovery, evolutionists announced to the world that Neanderthal man could not stand fully erect and that he walked with a bend at the knees. Dobzhansky says the Neanderthals "were of short stature, about 5 feet, but of exceedingly rugged build. They had a stooping posture, powerful neck muscles, and massive heads with surprisingly large brains—1450 cubic centimeters on the average, that is, as large as or larger than in modern man."[24] Statues appeared in museums portraying the Neanderthal man as stoop-shouldered and standing in a semi-erect position. An expert witness in the Scopes Trial, Fay Cooper-Cole, testified that the Neanderthal man's knees were habitually bent and that he walked in a semi-erect position.[25]

At last the evolutionists had uncovered the missing link, or so they claimed. As time passed, however, more Neanderthal skeletons were found. Reconstruction showed that they stood upright and had features identical to those of modern man. Scientists re-examined the original skeleton and found that the famous "prehistoric man" was simply an old man suffering from arthritis.[26] Again, throughout the world, museums silently began to remove their statues.

The truly amazing feature of the Neanderthal skeletons is that the cranial capacity is larger than that of modern man's. How could modern man have evolved from Neanderthal man? More likely, modern man has degenerated from Neanderthal man. This accords perfectly with the Bible, which shows that man is degenerating. Once again, "fact" has become fiction.

Cro-Magnon Man
(Homo Sapiens)

The original fossil skeletons of Cro-Magnon man were

found at Gower, Wales, and Aurignac, France. At Aurignac, seventeen skeletons were found. They resembled modern man so closely that they were buried in the village cemetery and thus lost to science for some time.[27] It was not until 1868 that five more skeletons were found at Cro-Magnon, France. These five skeletons established the Cro-Magnon type.

Amazingly, these fossil men are identical to modern man with one exception: their cranial capacity is greater than ours. As evolutionists Dodson and Dodson explain,

> *Around a hundred of these fossils, called Cro-Magnon after the French cave in which they were first found, are now known. . . . Their characteristics are well established and it may be said that they did not differ significantly from many modern Europeans. The cranial capacity appears to have been somewhat greater.*[28]

Regarding the culture of Cro-Magnon man, evolutionist John Waechter comments: "The food supply seems to have been more than sufficient, and the development of various forms of art in both eastern and western Europe suggests that Cro-Magnon Man had an advanced culture and a great deal of leisure time."[29] Waechter further states that the Cro-Magnon men were clearly modern in every way.

It is obvious that Cro-Magnon man looked like modern man and had a culture like that of modern man. Where, then, is the argument for evolution? Cro-Magnon man was a modern man, except that modern man has degenerated a little more than Cro-Magnon man.

Conclusion

In conclusion, all so-called "prehistoric men" can be placed into one of three categories: (1) extinct apes or monkeys, (2) living apes or monkeys, and (3) modern man. According to Merrell, Neanderthal man and Cro-Magnon man existed at the same time, with the latter replacing the former:

> *These tall and well built men, of the so called Cro-Magnon type, had a distinct bony chin on the front of the jaw, a high domed, thin walled skull, and greatly reduced brow ridges, and are indistinguishable from modern men. These are the people who, in a relatively short time, completely replaced the Neanderthal type. However, just where this modern type of man came from and who his immediate predecessors were is far from clear.*[30]

Perhaps Neanderthal man represented the line of Cain and the seemingly more advanced Cro-Magnon man represented the righteous line of Seth. Certainly, the Flood should have fossilized some antediluvian men.

The antediluvian civilization was an advanced civilization. These people invented musical instruments, including stringed instruments and wind instruments (Genesis 4:21). Many of them dwelt in cities (Genesis 4:17). Some scientists depict the antediluvian people as primitive, savage, cave dwellers, but the scriptural evidence refutes this idea.

Some people probably did live in caves prior to the Flood and there is some fossil evidence to indicate this, but people also dwelt in cities. It is possible that some archaeological excavations of ancient cities and civilizations are the remnants of the antediluvian world. Even today some people live in caves and are very primitive, but this in itself does not make them prehistoric. These people simply became isolated from mainstream civilization and declined to their present condition.

Even from evolutionists' viewpoint the theory of human evolution faces many unresolved difficulties. As Washburn stated in *The Evolution of Man,* "Some of the oldest and most troublesome questions about human evolution remain unanswered."[31]

Footnotes
Chapter Four

[1]Peter Stoler, p. 77.

[2]Merrell, p. 340.

[3]Dobzhansky, p. 325.

[4]Sherwood L. Washburn, "The Evolution of Man," *Scientific American*, Vol. 239, No. 3 (Sept. 1978), p. 208.

[5]Timothy Prout, "Has Evolution Stopped?", *Science Digest*, Vol. 48. No. 1 (July, 1960), p. 68.

[6]Merrell, p. 333.

[7]Horatio Hackett Newman, *Evolution, Genetics and Eugenics*, 3rd ed. (Chicago: University of Chicago Press, 1932), p. 155.

[8]Merrell, pp. 340-341.

[9]*Ibid.*

[10]Dodson and Dodson, 2nd ed. p. 246.

[11]Bolton Davidheiser, *Evolution and Christian Faith* (Grand Rapids: Baker Book House, 1969), p. 347.

[12]H. H. Wilder, *The Pedigree of the Human Race* (Henry Holt and Co., 1926).

[13]Newman, p. 150.

[14]Duane T. Gish, *Evolution—The Fossils Say No* (Creation Life Publishers, 1973), p. 64.

[15]Ashley Montagu, *Man—His First Million Years* (World Publishing Company, 1957), p. 64.

[16]Stoler, p. 64.

[17]Dobshansky, p. 326.

[18]Dodson and Dodson, p. 241.

[19]*Ibid.*, p. 239.

[20]*Ibid.*

[21]*Ibid.*, pp. 238-239.

[22]*Ibid.*, p. 247.

[23]Newman, pp. 154-155.

[24]Dobzhansky, p. 331.

[25]Davidheiser, p. 92.

[26]James A. Michener, p. 248.

[27]Newman, p. 156.

[28]Dodson and Dodson, p. 251.

[29]John Waechter, *Prehistoric Man* (Octopus Books, 1977), p. 186.

[30]Merrell, p. 338.

[31]Washburn, p. 208.

Chapter 5

EVOLUTION—A PHILOSOPHY

True science is the knowledge of facts and natural laws obtained through experimentation, measurable observation, and precise testing. When science goes beyond these realms, it enters the areas of speculation and philosophy. Evolutionist Horatio Newman acknowledges that the theory of evolution is a philosophy:

> *Let no one imagine, as he is conducted by the materialistic scientist in the paths of evolution from the inorganic to the organic, from the organic to the animate, from the animate to the rational and moral, until he lands, as it seems to him, logically and inevitably, in universal materialism—let no such one imagine that he has walked all the way in the domain of science. He has stepped across the boundary into the domain*

of philosophy.[1]

The world-renowned evolutionist Julian Huxley commented, "I find myself inevitably driven to use the language of religion. For the fact is that all this does add up to something in the nature of religion: perhaps one might call it evolutionary humanism."[2]

Furthermore, belief in evolution necessitates the acceptance of certain assumptions that cannot be experimentally verified. G. A. Kerkut, though an evolutionist himself, lists seven basic assumptions demanded by the presently accepted theory of evolution:

1. *Non-living things gave rise to living material, i.e. spontaneous generation occurred.*
2. *Spontaneous generation occurred only once.*
3. *Viruses, bacteria, plants, and animals are all interrelated.*
4. *The protozoa (single-celled organisms) gave rise to the metazoa (multi-celled organisms).*
5. *The various invertebrate phyla are interrelated.*
6. *The invertebrates gave rise to the vertebrates.*
7. *Within the vertebrates, the fish gave rise to the amphibia, the amphibia to the reptiles, and the reptiles to the birds and mammals.*[3]

Despite these unproven assumptions, many scientists boldly proclaim evolution as a fact of science and no longer a theory, as the following examples demonstrate.

> **George Simpson:** *Evolution is a fact and. . .we ourselves and everything in the world about us are products of an evolutionary history.*[4]
>
> **Roy Gallent:** *However, today there is hardly a scientist alive who does not accept as fact the idea of evolution through natural selection. But there are religious groups that still refuse to accept evolution, in spite of the overwhelming evidence supporting it.*[5]
>
> **Horatio Newman:** *Some of the weaker elements of*

the theory have given way under stress, and the whole selection factor as a primary causal factor in evolution has been seriously called into question; but since Darwin's time the fact of evolution has been almost universally accepted. Nevertheless, he was the first to furnish a really adequate proof of the fact of evolution.[6]

David Merrell: *It has become routine for biologists to preface their remarks about evolution with the statement, "Everyone now accepts the fact of evolution."*[7]

Sir Julian Huxley: *The first point to make about Darwin's theory is it is no longer a theory, but a fact. No serious scientist would deny the fact that evolution has occurred, just as he would not deny the fact that the earth goes around the sun.*[8]

Dodson and Dodson: *Fragmentary though the fossil record is, it gives striking testimony to the fact of evolution, and considerable detail can be worked out for many lines of descent.*[9]

When people accept the theory of evolution as fact, evolutionists do not have to cope with the enormous number of unanswerable questions regarding it. Thus they avoid many embarrassing dilemmas. Statements from evolutionists, however, do not automatically make evolution a fact.

Creation and evolution are both theories of origin. Since neither creation nor evolution can be subjected to the scientific method, the choice of theories becomes a matter of faith. This places both theories into the area of religion. The creationist has faith in God's ability to create all things, whereas the evolutionist has faith in nature's ability to evolve into all things. The choice is between faith in the divinely-inspired Word of an all-knowing God or faith in the reasoning, assumptions, and even guesses of man.

The Natural Scientist

We owe much to the natural scientist. He has protected us from many dreaded diseases and provided us with wonderful inventions to make life easier and more enjoyable. This influences many people to look to science for answers to all questions concerning life, including its origins. This leads into philosophy and religion, areas in which the natural scientist is not qualified to speak authoritatively because these are not subject to analysis by the scientific method. Nevertheless, the scientist attempts to answer all the questions. This results in theories science cannot prove, such as evolution and uniformitarianism.

Two Basic Philosophies

With respect to the question of origins, there are two basic philosophies: creation and evolution, with several variations of each. Scientific analysis cannot prove or disprove either one of these philosophies because no human was present when the earth came into existence. Therefore, the theories of origin do not belong in the realm of science, but rather in the domain of philosophy or religion. Each individual must examine all of the available data and choose the philosophy that best fits the evidence. If every scientist and layman were to do this, the theory of evolution would fall into the category of disproved theories.

When one deals with the question of origins, he immediately goes from science to speculation. Certainly science should attempt some explanation of origins; otherwise science would have no sense of direction. But who can prove by the scientific method what took place thousands, millions, or even billions of years ago? What scientific experiment can one conduct in the laboratory to prove beyond question what happened in the distant past?

Who can step back into the past before any man set foot on the face of the earth? Of course, no one can, even though evolutionists would like to make us believe otherwise. For example, Julian Huxley presumes to go back in time, using evolutionists' interpretation of the fossil record: "Before a million years ago we find no men; much earlier, no reptiles, then earlier again, no land forms at all, but fish as dominant vertebrates; and finally, no vertebrates with jaws or true paired limbs, but only jawless, limbless animals related to the lampreys."[10]

God challenged Job, "Where wast thou when I laid the foundations of the earth? declare, if thou hast understanding" (Job 38:4). Similarly, we must ask ourselves a challenging question: "Where was the evolutionist when the foundations of the earth were laid?"

We must place the question of origins, therefore, in the realm of religion. Whether one believes in creation or in evolution, he must exercise faith. For the creationist it is faith in God—a God who has all power, controls the universe, and has the ability to create matter, even man, from nothing. (See Hebrews 11:3; Genesis 1:1; 2:7.) For the evolutionist, it is faith in matter and the ability of matter to reorganize itself innumerable times during millions and billions of years. Either God created matter, or matter brought itself into existence out of nothingness. Either God created life, or life came out of non-living matter.

Which theory—evolution or creation—requires more faith? Is it harder to believe in an omnipotent, omniscient God, or to believe in the creative power of non-living, non-reasoning matter? The evolutionist may laugh at and mock the Christian's faith in God, but the religion of evolution demands even more faith than that of the creationist. Evolutionists sneer at the miracles of the creation account because no one can verify them experimentally, yet they adhere to evolution which no one can verify by the scientific method either.

Many Scientists Do Not Examine The Facts

Many scientists believe in evolution simply on the basis of tradition. They believe the theory of evolution simply because they know others believe it, yet they never research the so-called evidences of evolution for themselves. They accept the teachings of other scientists as fact without examining the evidence. Most scientists who believe in the evolutionary theory of origins have only a surface knowledge of the theory and accept the rest as true because other scientists have said it is so. Many of them have never studied Charles Darwin's book, *The Origin of Species*, which is the bible of evolution. They merely accept what others tell them about it. Their claim of faith in evolution is as unjustifiable as those who claim to have faith in God, yet just accept the traditional teaching of others without studying the Bible for themselves.

Today, educated men and women are recognizing evolution for what it is—a philosophy that merely interprets facts of science through an atheistic approach. As evolutionist Newman remarks:

> *Reluctant as he may be to admit it, honesty compels the evolutionist to admit that there is no absolute*

> *proof of organic evolution. The theory of geologic evolution meets with scarcely any opposition today, although its foundations are no more securely based than are those of organic evolution.*[11]

History of Evolutionary Thought

Most evolutionary texts imply that the theory of evolution is a product of modern-day science and began with Charles Darwin, but that creation originated from the ignorant superstitions of ancient man. In actuality, evolution is not a modern concept, nor did it begin with Charles Darwin. Evolution, though not termed as such, has been a common philosophy since the earliest recorded history of man.

Ancient Greek philosophers such as Anaximander and Empedocles taught that men evolved from fish and animals came from plants.[12] This was typical of many ancient thinkers, as Paul Moody notes: "Ideas that by one means or another evolution does occur far antedated Darwin, however. In fact, such ideas are probably as old as human thought."[13]

Most evolutionists, in defense of their theory, state that man believed in some form of supernatural creation due to superstitions and ignorance, as Dunbar and Waage express:

> *To early men who knew at most a few hundred kinds of animals and plants and had only superstition to guide them, it seemed reasonable to believe that the particular god they feared and worshipped "molded each species out of the dust of the earth and breathed into it the breath of life." The story is much the same among primitive peoples throughout the world.*[14]

Although the ancients did produce many superstitions concerning creation, there are just as many superstitions about strange methods and forms of evolution. The very argument used by the evolutionists to support the theory of evolution actually discredits it.

Charles Darwin and the 1800's

Charles Darwin's publication in 1859 of *The Origin of Species by Means of Natural Selection* did not cause an unexpected, sudden appearance of the theory of evolution. In reality, many people in the early 1800's were suggesting that life might be the result of evolution. "The possibility that the world of life might be a product of evolution was, indeed, 'in the air' when Charles Darwin started his work."[15]

In 1809, the same year that Charles Darwin was born, Jean Baptiste Lamarck published a book entitled *Philosophie Zoologique*. This book contained the first scientifically complete theory of evolution. The mechanism of evolution that Lamarck advocated was the inheritance of acquired characteristics. Darwin, though not willing to admit it, was strongly influenced by Lamarckian theory.

Although there were many advocates of evolution prior to Darwin, he must be credited with causing the acceptance of the theory by a large percentage of the scientific public during the latter part of the 1800's.[16] The reasons for the acceptance of Darwin's theory were not all scientifically based, for many scientists, according to evolutionist Sir Julian Huxley, were simply "ready to be persuaded."[17]

Charles Darwin was born on February 12, 1809, in Shrewsbury, England. As a young man, Darwin attended Christ's College in Cambridge where he completed the requirements for a degree in theology. This was the only degree Darwin ever received, but scientists do not try to discredit his theory of evolution for this reason although they often ridicule theologians who defend creationism.

After the completion of his degree, Darwin accepted an invitation to become a naturalist on the voyage of the *Beagle*. On this voyage, Darwin first began to develop his concept of natural selection, which is the mechanism that he proposed for evolution in his book, *The Origin of Species*.

Many historians portray Darwin as totally without outside

influence while he was on the *Beagle* developing his "natural selection" theory, but actually Sir Charles Lyell's *Principles of Geology* had a great impact on him.

> *By converting the young Darwin to the belief that the earth was formed by natural causes,* Principles *in large measure paved the way for Darwin's theories on evolution. In fact, Darwin once confessed feeling that his books "came half out of Sir Charles Lyell's brain."*[18]

In fact, the book had such a profound effect upon Darwin that he dedicated the journal of his voyage to Lyell: "To Charles Lyell, esq., F.R.S., this second edition is dedicated with grateful pleasure, as an acknowledgment that the chief part of whatever scientific merit this journal and other works of the author may possess, has been derived from studying the well-known, admirable *Principles of Geology*."[19]

In short, the theory of evolution was not the result of one man, Charles Darwin, though he must be credited with its general acceptance. The theory resulted from a group of people who were willing to discard or destroy some of the principle teachings of the Bible.

Footnotes
Chapter Five

[1]Newman, p. 548.

[2]Julian Huxley, rev. ed., p. 132.

[3]G. A. Kerkut, *Implications of Evolution* (Pergamon Press, 1960), pp. 6-7.

[4]Simpson, p. 169.

[5]Gallent, p. 4.

[6]Newman, p. 548.

[7]Merrell, p. VI.

[8]Sol Tax and Charles Callender, *Issues In Evolution* (University of Chicago Press, 1960), p. 41.

[9]Dodson and Dodson, pp. 85-86.

[10]Julian S. Huxley, 1st ed., p. 27.

[11]Newman, pp. 49-50.

[12]Richard E. Leakey, "Skull 1470," *National Geographic*, Vol. 143, No. 6 (June 1973), p. 75.

[13]Peter Stoler, p. 3.

[14]Dunbar and Waage, p. 109.

[15]Dobzhansky, p. 357.

[16]Gallent, *Charles Darwin*, p. 12.

[17]Sol Tax, *Evolution After Darwin*, (The University of Chicago Centennial), p. 11.

[18]Hotton, p. 11.

[19]Newman, p. 7.

Chapter 6
THE UNIFORMITARIAN ASSUMPTION

Evolutionists assume the doctrine of uniformitarianism, and evolution depends heavily on this assumption. Uniformitarianism means that existing processes occurring at their present rate and manner are sufficient to explain geological history. By studying present processes we can arrrive at an adequate understanding of earth's past. In other words, the present is the key to the past.

Uniformitarianism versus Catastrophism

One of the greatest areas of disagreement between the creationist and the evolutionist involves the concepts of catastrophism and uniformitarianism. Most creationists strongly advocate catastrophism, whereas most evolutionists strongly advocate uniformitarianism.

Horatio Newman defines and contrasts these terms as follows:

> *The prevailing theories as to historical geology during the middle ages involved the idea of "catastrophism." According to this view all important changes in the earth's crust represented sudden radical transformation, involving earthquakes, volcanic outbursts, floods, sudden upliftings of submerged areas, or equally sudden submergence of land bodies. The Scotch philosopher and geologist, Hutton, who lived during the last half of the Eighteenth Century, combatted the idea of catastrophism by advocating the doctrine of "uniformitarianism," a view involving the idea that past changes on the earth were the result of the same sort of gradual changes as are observed to be taking place today—in brief, that there has been a strict uniformity of change throughout the entire period of geologic history. There may have been, according to this view, local catastrophies, such as volcanic outbursts, earthquakes, and floods, but the main trend of change has been slow and constant, due largely to erosion and allied facts.*[1]

Catastrophism pre-dated uniformitarianism; prior to Darwin's time catastrophism was the predominant scientific theory used to explain geological facts.[2]

One of the major reasons for the decline of catastrophism in the 1800's was that it did not allow for the immense time required for evolution. Many people in the early 1800's were suggesting the possibility of the evolution of life. In order to present a plausible theory of evolution, they needed a theory of earth's formation that allowed for a great expanse of time: "Charles Darwin needed as much time as possible for the evolution of the amazing array of life forms known by 1859;

millions of years seemed indicated before the development of man."[3]

Just prior to this time, James Hutton, a doctor and agriculturist, formally presented the principle of uniformity. According to this principle, the present processes of erosion, sedimentation, volcanism, glaciation, and so on continued at the same rate through millions and billions of years and finally produced all of the geological features now seen on the earth. With this idea, the evolutionist was able to provide the large amount of time that his theory demanded.

Hutton's theory of uniformity, presented in 1785, did not receive widespread acceptance at first and had practically no influence on the ideas of that time. When evolutionists needed a theory that would give them immense amounts of time, however, they readily accepted the theory of uniformitarianism. It is now the accepted theory of the majority of scientists and has been for approximately 120 years.

Let us now see whether geological findings support the uniformitarian concept. We should remember that uniformitarianism allows no major geological catastrophe; if we can identify one in geological history, then uniformitarianism is discredited. When we survey the evidence, we discover many such contradictions of uniformitarianism, the most obvious of which are the Columbia plateau, thrust faulting, and frozen mammoths.

The Columbia Plateau

The Columbia Plateau is an immense lava bed located in the northwestern part of the United States and covering large portions of a number of states. The lava bed has an amazing depth of 2500 feet and is spread over an area of 200,000 square miles. Dunbar and Waage attempt to explain its formation through a gradual process, thus reconciling it to uniformitarianism:

> *The early basalt flows, seeking the lowest places,*

gradually filled the basins and eventually buried the hills to a depth of some 2500 feet. As the flows spread over an area of some 200,000 square miles they interrupted the drainage, damming the streams and giving rise to lakes and swamps around the margins of the province in which sedimentary deposits interbedded with the lavas include abundant and well-preserved plant fossils.[4]

If the present is the key to the past as uniformitarianism states, then why do we not see lava beds of the magnitude of the Columbia Plateau forming somewhere on the earth today? Also, how can the unformitarian explain the vastness of such lava flows in comparison with the small volcanic upheavals and lava flows presently occurring on the globe?

There is only one reasonable explanation for the formation of the Columbia Plateau—a sudden catastrophic volcanic upheaval. Dunbar and Waage admit that the Masa Basalt, one of the lava flows making up the Columbia Plateau, now covers 100,000 square miles, including parts of three states. Certainly this does not indicate a gradual change but rather a sudden cataclysm. Evolutionists have no reasonable explanation for the formation of the Columbia Plateau, yet creationists can point to the Great Flood and the tremendous volcanic upheavals associated with it to explain such a phenomenon.

The Columbia Plateau is by no means the only example of immense volcanic activity and resulting lava flows. Volcanic plateaus are scattered across the globe. One such plateau in South America covers an area of 300,000 square miles. Even with the presentation of only one isolated example such as the Columbia Plateau, the "fact" of uniformitarianism becomes fiction.

Thrust Faulting

Found in most areas of the globe are stratified layers of rock or soil called "geologic columns." According to uniformitarianism, these layers were formed over millions or even billions of years through the processes of erosion, glaciation, deposition, and volcanism.

These processes have supposedly acted at the same rate throughout the geological history of the earth. According to this concept, the stratigraphic layers should become progressively older as one goes down the geologic column, with the youngest layer on top and the oldest on bottom. Contrary to this predicted arrangement, in many places throughout the globe, "old" layers are superimposed on top of "young" layers. Uniformitarians call this a "thrust" or "thrust fault."

We can accept thrust faulting on a small scale (such as a few hundred feet) and can explain it by natural processes, but many thrust faults encompass hundreds and even thousands of miles. A few examples are the Sevier Orogeny, Heart Mountain Thrusts, Lewis Overthrust, and the Matterhorn and Mythen Peaks in the Alps. Many of these thrusts yield no physical evidence (in the form of intense faulting and folding) that a thrust has even occurred.

Unfortunately for uniformitarianism, there is no known method by which such large blocks of rock could overcome frictional stress and be moved any distance while still remaining intact. Yet, many thrusts have displaced rock strata and moved them for miles. Many scientists consider this to be mechanically impossible, at least by natural processes known today. As an example, we will discuss the Sevier Orogeny in some detail.

The Sevier Orogeny

The Sevier Orogeny covers significant portions of Nevada, Utah, Idaho and Wyoming. It is in the shape of a massive belt, composed of supposedly older rock strata superimposed over supposedly younger rock strata. According to evolutionists:

> *The Sevier Orogenic belt formed along the east flank of the Cordilleran Geanticline and was characterized chiefly by great thrust sheets driven eastward as*

> *much as 25 miles, in places carrying rocks as old as Cambrian over the Cretaceous deposits. This great belt of Cretaceous low-angle thrusting extended from southeastern Nevada well into Idaho.*[5]

How is it possible for such an immense belt of rock matter to move 25 miles? Obviously, no known mechanism outside of catastrophism is capable of producing such a phenomenon as the Sevier Orogeny.

Frozen Mammoths

One of the most puzzling situations for uniformitarianism is the existence of numerous frozen mammoths in Siberia and Alaska. Dunbar and Waage describe the discovery of one such mammoth:

> *In August, 1900, a Russian hunter, following a wounded deer along the valley of Beresouka River in eastern Siberia, came upon the head of an elephant sticking out of the frozen ground. He chopped off a tusk, which he later sold to a Cossack through whom the news reached St. Petersburg that another frozen mammoth*

had been found. An expedition organized the The National Acadamy of Science to collect the specimen reached the locality in September, 1901, after a journey of some 3,000 miles, and after the back of the animal had been exposed to the warmth of two summers and part of the flesh had been gnawed away by wild animals. Excavation soon revealed, however, that the buried portion was still intact, and on one of the limbs the flesh was "dark red in color and looked as fresh as well-frozen beef or horse meat." Indeed, scraps of it were eagerly devoured by the collectors' dog team.[6]

The mammoths were frozen so rapidly that many were found with undigested grasses, bluebells, buttercups, tender sedges and wild beans yet in their stomachs and mouths.

The existence of the frozen mammoths poses many questions that uniformitarianism cannot answer. For example, how were the mammoths quick-frozen in seemingly the very spot they were grazing? How could the climate change so rapidly that at one moment plants such as buttercups, bluebells, and tender sedges were growing and in the next moment the entire area became a cold-storage locker? As Thomas Fleming observes, "Today the kind of vegetation they fed on does not exist in Siberia. Muck does not feel that anyone has been able to explain satisfactorily the sudden cold that 'quick-froze' these cousins of the elephant and turned Siberia into the giant icebox it is today."[7]

Obviously, only a cataclysm of some kind can account for sudden destruction and quick-freezing of the mammoths. Most probably the quick-freezing of the mammoths occurred during the sudden breakdown of the vapor canopy during the Flood, which would have caused a great drop in atmospheric pressure and temperature. Thus the Flood was probably the indirect agent of the cataclysm.

In summary, the present processes working today have molded the earth in the near past—that is, since the Flood—

and will continue to mold the earth in the near future. (See Genesis 8:22; II Peter 3:7.) Geological findings demonstrate, however, that catastrophic events occurred in earth's past. The subtle processes presently at work simply could not have molded earth into its present condition.

Footnotes
Chapter Six

[1]Newman, pp. 22-23.

[2]Dott and Batten, p. 22.

[3]*Ibid.*

[4]Dunbar and Waage, p. 425.

[5]*Ibid.*, p. 373.

[6]*Ibid.*, p. 34.

[7]Thomas Fleming, "Solving the Lost Continent Mystery," *Reader's Digest*, Vol. 113, No. 676 (Aug. 1978), pp. 132-133.

Chapter 7

SPONTANEOUS GENERATION

The phrase "spontaneous generation of life" simply means that life came into being from lifeless (inorganic) matter by forces which are inherent within matter itself. Before the 1600's, spontaneous generation was the dominant consensus of the scientific public.[1] Men such as Empedocles in the fifth century B. C. and Anaximander of the sixth century believed strongly in the possibility of spontaneous generation of life from non-living materials. They thus settled, in a very simple fashion, the question of the origin of life.[2]

According to primitive theories of spontaneous generation, horsehairs could change into living snakes in rain water, filthy substances could change into frogs, rags could change into rats, or decaying meat could become living maggots.[3] Hotton describes how this theory apparently arose:

> *In ancient times such phenomena as the appearance of frogs in great numbers following spring floods along the Nile and the appearance of maggots in decaying meat had given rise to the notion that under certain circumstances life could arise from nonliving matter. The concept (it became known as the theory of "spontaneous generation") obviously resulted from insufficient attention to external phenomena.*[4]

Clearly, then, the concept of spontaneous generation has been with man for millenia.

Francisco Redi

Scientists before the 1600's sincerely thought they had solid evidence to support the theory of "spontaneous generation of life." For example, everyone knew that dirty rags left unattended in a closet for a long enough period of time would become infested with rat nests. To the early scientist this was definite evidence that rats came from dirty rags.

In 1668, however, an Italian scientist named Francisco Redi dealt a crushing blow to the theory of the spontaneous generation of life. He conducted a simple but convincing experiment to disprove that flies arose spontaneously from decaying meat.[5] Redi placed a piece of meat into three separate widemouthed flasks. He left one flask open to the air, covered the second with a porous cloth, and covered the third with paper.

The meat in all three flasks decayed, and as the meat decayed its odor attracted flies. The flies entered the open flask and laid eggs on the decaying meat. The eggs soon hatched into maggots, a stage in the life cycle of the fly between the egg and adult.

The odor from the flask covered by a porous cloth also attracted flies, which laid eggs on the cloth. These eggs also hatched into maggots but, due to the cloth that formed a barrier, they could not get to the meat. In the third flask, the paper covering protected the meat by forming an odor barrier. No maggots appeared in the meat or on the paper covering.

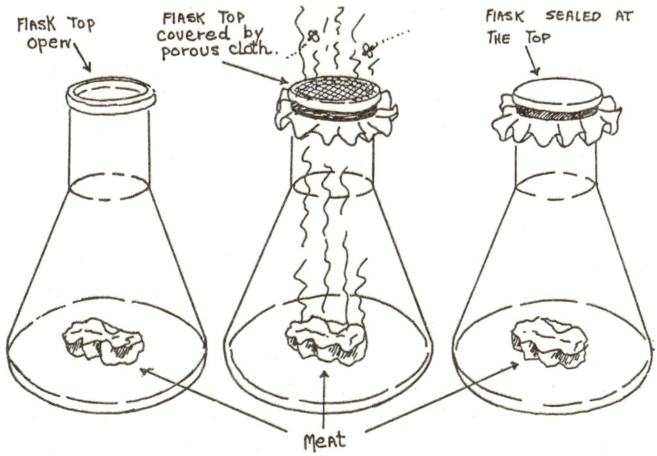

Redi thus demonstrated one of the basic properties of life: life comes only from other life. Scientists now call this the Law of Biogenesis. This is, of course, what the Bible teaches in Genesis 1:11-12, 1:21, and 1:24-25.

Louis Pasteur

After the demonstration by Redi, the consensus of the scientific world concerning the spontaneous generation of large creatures changed. It became generally accepted that larger creatures required parents to produce the offspring. Most scientists still maintained, however, that the amazing microscopic world (bacteria, algae, etc.) could arise spontaneously from non-living matter. This concept of the microscopic world was upheld until the mid-1800's. We should note that scientists supported this theory of spontaneous generation with much so-called factual evidence:

> *There appeared champions for and challengers of the theory that living things can originate spontaneously, each with a new and sometimes fantastic explanation or bit of experimental evidence. In 1749 John Needham (1713-1781), experimenting with meat exposed to hot ashes, observed the appearance of organisms not present at the start of the experiment and concluded that bacteria originated from the meat.[6] The last major advocate of spontaneous generation was Pouchet, who published in 1859 an extensive report "proving" its occurrence.[7]*

The deathblow to the theory of spontaneous generation came in 1864. Louis Pasteur demonstrated by experiment that microscopic life comes only from other microscopic life. He conducted his experiment by first sterilizing some bacterial broth in a flask. He then sealed the flask. After a period of time, no bacteria formed in the broth.

Some scientists, however, opposed the experiment on the grounds that air, which was sealed out of the flask, was necessary to produce life in the broth. Pasteur conducted the experiment again, except this time he placed the sterilized broth in a flask with a narrow top bent into the shape of an "S."[8] The flask allowed fresh air to come in contact with the broth, yet trapped any dust or particles carrying bacteria in the bend of the neck of the flask. Again no bacteria formed in the broth, thus demonstrating that life comes only from other life.

Pasteur conducted this experiment over 100 years ago, and the original flasks of broth are on display today at the Pasteur Institute in France. They are still free of bacteria.

It seems that scientists would have discarded once and for all the theory of spontaneous generation, but it has gained new life through the modern theory of evolution. Merrell admits that evolution incorporates a new version of the old disproved theory of spontaneous generation:

> *In recent years, a new attack has been made on the problem, and the result has been, interestingly enough, a new version of spontaneous generation. The theory proposes that life originated on earth in the past when conditions were different from those of the present, and was preceded by a gradual chemical evolution that ultimately gave rise to self-duplicating molecules.*[9]

Evolutionists have made certain, however, that a modern-day scientist cannot disprove the newly advocated theory of spontaneous generation. They have made it so hypothetical, and buried it so far in the past, that no scientist can possibly use the scientific method to prove or disprove the theory. In fact, most evolutionists contend that the spontaneous generation of life must have occurred under conditions which we cannot obtain today and at which we can guess vaguely.[10]

Three Alternatives

How did life on this planet begin? This question has arisen time and time again as we look across the pages of human history. After years of intense study, research, and compilation of data, scientists have narrowed the possible options regarding the origin of life to three: (1) in the beginning, God created; (2) in the beginning, life was carried to the earth from some other part of the universe; (3) in the beginning, matter evolved.[11]

Most evolutionists reject the option of a supernatural creation. The second option—often called the "pansperm hypothesis" or the "cosmozoic theory"—is not very attractive either:

> *A second theory of the origin of life, the cosmozoic theory, is that the original spores of life reached the earth accidentally from some other part of the universe. This theory is completely unsatisfactory for two reasons. First, because of the intense cold, extreme dryness, and the intense radiation of interstellar space, the probability that even the most resistant living spores could withstand exposure to interstellar space is vanishingly small. Second, the theory does not explain the origin of life at all, but merely changes the scene of origin from the earth to some remote and undefined part of the universe.*[12]

Even if the idea of spores floating in from other planets were correct, it merely transfers the basic question of life's beginning to some other place.[13] Let us therefore take an in-depth look at the only remaining option available to the atheistic evolutionists, the spontaneous production of living matter from non-living matter.

One of the major questions debated amongst evolutionists is how non-living matter evolved into living matter. This question is more significant than most people realize, for if the evolutionist cannot demonstrate or at least provide a supporting theory as to how the first living matter was spontaneously generated, then the remainder of the theory is in grave jeopardy.

Leading evolutionists have written many books and articles in an attempt to resolve the dilemma of how life began, but apparently one evolutionist's guess is as good as another's. Evolutionists do agree, however, that there is a large gap separating the non-living world from the living. As evolutionist Richard Dickerson states:

> *The step from aldehydes and amino acids non-biologically formed to a living cell is a giant one. It is one thing to propose scenarios for the origin of life that might have been; it is another thing entirely to demonstrate that such scenarios are either possible or probable.*[14]

The Primordial Atmosphere

Most evolutionists postulate a primordial atmosphere having the necessary constituents for the production of innumerable amino acids. According to them, this prehistoric atmosphere contained primarily methane, ammonia, carbon dioxide and water but lacked free oxygen. The postulated primitive atmosphere contained the exact substances needed to produce amino acids, yet no evidence whatsoever supports the postulate. It is certainly permissible to develop a postulate within the realm of science, but the postulate should at least have some supporting evidence.

The evolutionist is quick to label the creationist as unscientific for postulating different atmospheric and environmental factors before the Flood, even though the Bible—which has never been disproven—supports this view. Yet, the evolutionist proposes a primordial atmosphere totally different from that of today without any scientific basis except his faith in the doctrine of evolution.

Let us now look at the primordial atmosphere suggested by the evolutionist.

The proposal of a primordial atmosphere without oxygen in reality offers more questions than it supposedly answers. In a brief summation, here is the dilemma: an oxygen-free atmosphere had to be present during the vast expanse of time required for inorganic molecules to react with each other to produce bio-organic molecules, which are the precursors of life.[15] This is because bio-organic molecules are very unstable and readily break down in an oxidizing atmosphere such as the one today. As Dickerson explains:

> *Assuming that terrestrial life did evolve on the earth, what was the planet like when the process began? One thing is certain: the atmosphere contained little or no free oxygen and hence was not strongly oxidizing as it is today. The organic matter that must accumulate as the raw materials from which life could evolve is not stable in an oxidizing atmosphere.* [16]

Without free oxygen present, however, there would be no ozone layer.[17] (See Chapter 1.) Without the ozone layer to filter out deadly ultraviolet radiation, life on earth cannot exist. The absence of oxygen required to produce the precursors of life would have meant the immediate destruction of any life that formed.

There is no solution to this dilemma. The evolutionist has stumbled over the first step.

In summary, the formation of a primordial atmosphere is purely speculative. No one knows the exact composition of the early atmosphere; therefore, when scientists attempt to simulate the early atmospheric condition, they are merely speculating. Even if scientists should someday synthesize life from inorganic materials in a test tube—though they are nowhere close to doing so and doubtless never could do so—this still could not prove that life did indeed evolve. As far as the primordial atmosphere is concerned, this would not prove that the simulated laboratory conditions were the actual primordial atmospheric conditions.

This would only prove that it took the organized knowledge and intelligence of countless numbers of scientists over many years plus the most sophisticated equipment of modern technology to produce life. Even if scientists did synthesize life after such an expenditure of time, energy, knowledge, planning, and preparation, it would take a great leap of faith to conclude that non-living matter did this spontaneously without any outside help.

Of course, scientists have synthesized in the laboratory many basic chemicals used in living systems. Examples are sugars, polypeptides, nucleotides, and ATP. Scientists have not even come close, however, to creating a self-reproducing organism, not even a simple, one-celled amoeba. The vast difference is comparable to a few bolts and nuts in relation to a computer.

The basic question is this: Could blind matter, which cannot reason, think, understand, or plan, have produced such a highly complex world that even the most intelligent scientists and the most advanced computers cannot duplicate the first step in the process, or did an all-knowing God prepare and create our world?

Reactions In and Out of Water

Most evolutionists picture the primordial earth as a superheated molten mass that over millions of years gradually began to cool on the surface. As the earth cooled, water began to collect in pools and basins which later formed the oceans. Dodson and Dodson describe this cooling process.

> *A large quantity of the hydrogen and oxygen were probably united as water, but for long ages it was certainly present only as superheated steam. The hot vapors would rise toward the cold outer layers of the atmosphere, condense and fall as rain, only to be converted to steam again before striking the earth. Gradually the earth cooled sufficiently to permit the rainfall to strike the earth, then to begin to form pools and larger bodies of water. Optimum conditions for solubility and reaction existed, and the entire earth was a great crucible for random compound formation and reformation.*[18]

At first this seems advantageous for evolution, because the chemical reactions required to produce the initial precursors of life would have to take place in water.

To produce the next step, however, the reactions would have to take place out of water. The very water needed to begin the reaction would in the next step halt the formation of life. The water molecules would short-circuit the desired polymerization reaction since the presence of water favors depolymerization rather than polymerization.[19] How does the evolutionist get around this? Simply by using his imagination, as exemplified by Dickerson's explanation:

> *The difficulty of preventing competition by water molecules in a coupling reaction has encouraged biochemists to think of ways in which the amount of water in the vicinity of the polymerizing species might be reduced. An obvious possibility is evaporation, and one can "imagine" (emphasis added) a small portion of the Haldane Soup being concentrated by the solar evaporation of a tide pool near some Archean beach.*[20]

How can the imagination of scientists be reconciled with the scientific method? This is a typical example of the way evolutionists must revert to secondary bracing theories to prop up and support their crumbling primary theory.

The Living Cell

For many years scientists advocated the concept of a "simple cell." The lack of complexity of the "simple cell" was presented as evidence for evolution. After all, it was not too hard to visualize organic compounds floating in some ancient pool suddenly transforming themselves into some form of simple cell. Unfortunately for the evolutionist, it is now evident that any living cell is in reality a highly complex entity.

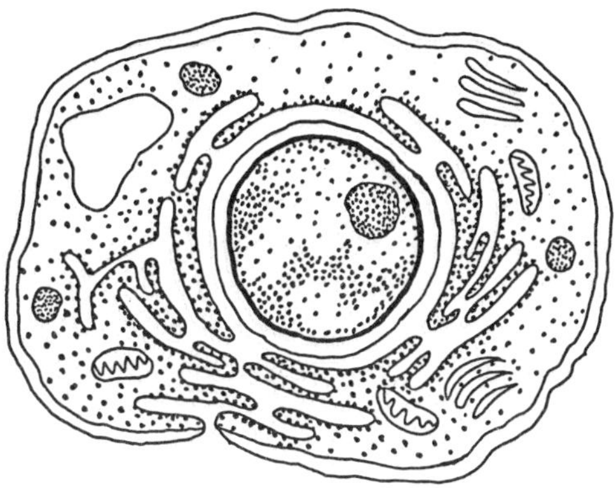

Cell

For example, according to Fuller and Carothers, a plant cell contains: (1) cytoplasm, which is composed of cytoplasmic membranes, cytoplasmic ground substance, plastids such as chromoplasts and leucoplasts, mitochondria and ribosomes; (2) a nucleus, which is composed of nuclear membrane, nuclear sap, nucleolus, and chromatin network; (3) ergastic substances, which include cell sap, starch grains, crystals, oil drops, and protein grains; and (4) a cell wall.[21] Space will not allow for a detailed discussion of these components. Suffice it to say that each is a detailed and complex chemical entity.

As the complexity of the living cell became apparent, evolutionists realized that their steppingstone to evolution had become a stumbling block. In fact it is a stumbling block of such immensity that the postulated evolutionary principles could never overcome it. For example, Julian Huxley describes the complexity of bacteria, the microscopic single-celled organisms often considered to be the most primitive living cells:

> *It used to be supposed, until very recently, that the whole enormous groups of bacteria were so primitive that they possessed neither chromosomes nor sex. One of the most spectacular discoveries of the past few years is the fact that, in their chromosomes and their mechanisms of sexual recombination, they possess essentially the same genetic equipment as you or I or any other organism.*[22]

Most evolutionists realize the difficulty of producing a plausible theory for the spontaneous development of a living cell. Many take the easy road out by simply relying on "Father Time." For example, Hotton admits he can offer no explanation for the necessary developments, but then argues that given enough time they could happen:

> *At what point were large coacervate globules able to break into smaller ones, each similar to the parent globule? These questions cannot be answered at this point and are perhaps unanswerable. But with a billion years of random combinations in which advantages were conferred by increasing complexity, it is not surprising that definitive features of living organisms, such as cell membrane and the cycle of recombination that we call reproduction, should have reappeared.*[23]

Other evolutionists argue that the first living cell was not as complex as the single cell organisms of today and that the creationists are unfair to point to their complexity. According to their own theory, however, the present is the key to the past! (See Chapter 6.) Furthermore, just what portion of the simplest present-day cell could we eliminate? How could we make it less complex and yet able to survive and reproduce? The answer is that nothing can be taken away. Every part of the living cell is vital to its existence.

Another insurmountable problem for the evolutionists is this: If the protozoa (single-celled organisms) are as simple as evolutionists imply, why are they not constantly changing? When one considers the countless billions of protozoa that live in every conceivable niche and corner of this planet, their extremely short life span, and their relatively simple and rapid method of reproduction, it seems that they would be the most likely candidates for evolutionary change. This in turn should make them so numerous in variety as to make them virtually impossible to classify. Scientists have classified the protozoa, however, just as they have classified higher plants and animals. This once more emphasizes that "like begets like," in accordance with Genesis 1:24-25.

Mathematical Probabilities of Spontaneous Generation

Evolutionists assert in their literature that "chance," operating over a long period of time, provided all the creative force needed for the spontaneous generation of life. Using this assumption, what is the mathematical probability of the chance appearance of protein-like molecules, the very basic building blocks of all living substances? Mathematicians in the field of probability have given a range of answers between 1 in 10^{123} to 1 in 10^{29345}.[24] (The scientific notation 10^n means 10 multiplied by itself "n" times; in other words, one followed by "n" zeros.) If this answer were taken out of scientific notation and typed on standard typing paper, it would take nearly twenty pages to give the answer.

These figures apply only to the spontaneous generation of protein-like molecules, which are actually non-living molecules. What about the mathematical probability of the spontaneous generation of the first living cell? According to mathematicians, the probability factor, even under ideal conditions, would be one chance in a number so large that if it were typed single-spaced on both sides of a page of typing paper, the mountain of typing paper needed would fill the entire solar system. Even the billions of years proposed by evolutionists do not begin to provide the time needed to make evolution probable from a statistical point of view.

If an open-minded person sincerely looks at these statistical facts, he will certainly relegate the theory of evolution to the realm of fantasy!

Could A Computer Factory Be Spontaneously Generated?

The spontaneous generation of a living cell is comparable to the spontaneous generation of the most highly complex computer conceivable by man, plus the factory that produced it, plus the power source to run it. The only problem with this comparison is that the factory and power source would have to be within the computer, rather than the computer being within the factory and the power supply being external.

Let us suppose—something the evolutionist likes to do—we could fill a gigantic box with wire, metal, plastic and the other basic building materials of a computer. Now let us shake the box for a billion years. Even after eliminating the destructive factors of friction, what is the probability that within a billion years a highly efficient computer would evolve? Not very good. Nevertheless, let us suppose by some process as yet unknown—a favorite assumption of the evolutionist—the computer did appear within the box. Although we would have a computer, it would not work because it would not possess a power supply. Therefore, the computer that took a billion years to generate spontaneously would be useless.

Let us really stretch our imaginations, and say that by some process as yet unknown the computer did contain its own internal power supply. This success is again only temporary, for after a few months the computer would wear out and cease to operate. For the success to be permanent, therefore, the computer must be able to manufacture duplicates that possess the same capabilities as the parent computer.

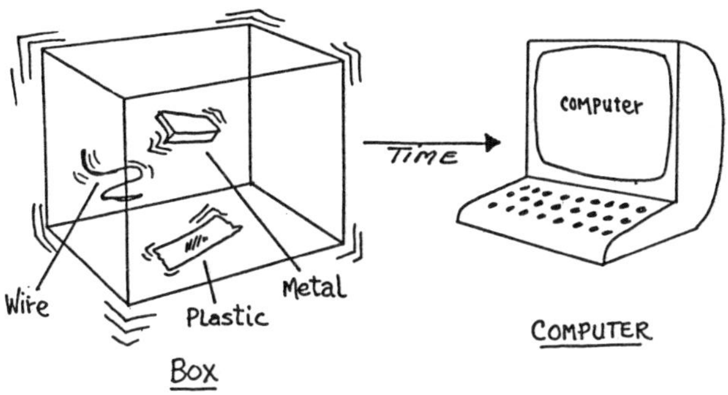

The above illustration may seem a bit elementary, but evolutionists recognize the same problems for the spontaneous development of a living cell as exemplified by our illustration. As Merrell admits:

> *The next question is the crux of the problem of the origin of life: How, from the dilute broth of organic compounds, did higher types of organization arise, persist, and ultimately lead to self-duplicating entities? Unfortunately, our knowledge here is only a beginning toward complete understanding.*[25]

Space does not allow for detailed discussion on the spontaneous generation of the first nucleus or for the first modes of reproduction. Suffice it to say that all so-called "facts" concerning these two areas are mere speculations on the part of evolutionists.

We conclude our discussion of spontaneous generation with evolutionist Dickerson's observations: "The evolution of the genetic machinery is the step for which there are no laboratory models; hence one can *speculate* (emphasis added) endlessly, unfettered by inconvenient facts."[26] When we eliminate the facts in this fashion, we are left only with fiction.

Footnotes
Chapter Seven

[1]Dobzhansky, p. 16.

[2]Merrell, pp. 14-15.

[3]Truman Moon, Otto, and Towle, *Modern Biology* (Holt, Rinehart, and Winston, 1960), p. 14.

[4]Hotton, p. 8.

[5]Holt, Rinehart, and Winston, pp. 13-14.

[6]Michael Pelczar and Reid, *Microbiology* (McGraw-Hill, 1958), p. 25.

[7]Pelczar and Reid, p. 27.

[8]Dodson and Dodson, pp. 106-107.

[9]Merrell, p. 59.

[10]Dobzhansky, p. 15.

[11]Huxley, p. 17.

[12]Dodson and Dodson, p. 107.

[13]Harlow Shapley, *Climate Change* (Harvard University Press, 1953), p. 8.

[14]Richard E. Dickerson, "Chemical Evolution and the Origin of Life," *Scientific American*, Vol. 239, No. 3 (Sept. 1978), p. 73.

[15]Dodson and Dodson, p. 109.

[16]Dickerson, pp. 75-80.

[17]*Ibid.*

[18]Dodson and Dodson, pp. 109-110.

[19]Dickerson, pp. 75-80.

[20]*Ibid.*, p. 80.

[21]Harry Fuller and Zane Carothers, *The Plant World*, 4th ed. (Holt, Rinehart, and Winston, Inc., n.d.), p. 46

[22]Huxley, p. 103.

[23]Hotton, p. 21.

[24]John Moore, *Questions and Answers on Creation/Evolution* (Baker Book House), p. 67.

[25]Merrell, p. 63.

[26]Dickerson, p. 85.

Chapter 8
DESIGN

When one looks into the starry sky on a dark night, he is immediately captivated by the beauty of design. From the vast expanses of the universe, which can be observed only superficially with the most powerful telescopes, to the infinitely minute microscopic world, which can barely be observed with the most powerful microscopes, everything points toward design.

Man with his limited knowledge is incapable of understanding the limitless God. In fact, man is even incapable of understanding all of God's creation. Who has not pondered the thought of a vast universe without boundaries or the infinitesimal particles that comprise matter? But man can observe that creation has design.

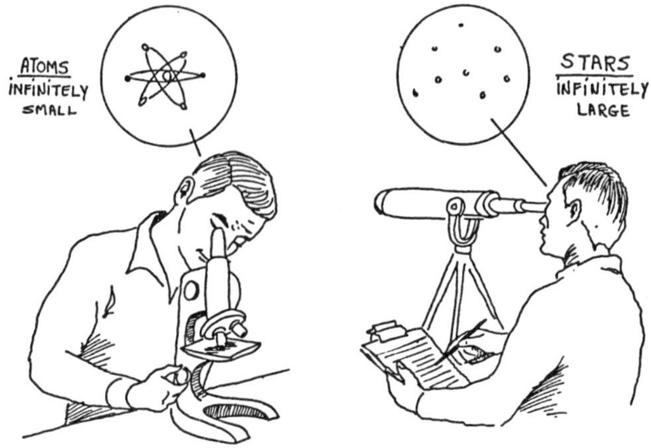

Everything is orderly; everything works in a set manner. When one looks at the organized complexity of a beehive or the millions of parts of the human body all working in unison, how is it possible to think of anything else but design?

Evolution, on the other hand, is the opposite of all evidence in nature. How could a random, haphazard, desultory process such as evolution produce the highly organized world of nature existing today?

Evolution requires much more faith than does creation. When we observe a building, common sense tells us a builder had to erect it. When we see a jet plane, we know a jet plane manufacturer had to assemble it. When we see a computer, we know a team of computer technicians designed it. How much more, when we look at the order of the universe, the complexity of the human body, or organization of a colony of bees, does common sense tell us there must be a Designer, a Creator?

Evolutionist Lewontin remarks, "Life forms are more than simple, multiple and diverse, however. Organisms fit remarkably well into the external world in which they live.

They have morphologies, physiologies, and behaviors that appear to have been carefully and artfully designed to enable each organism to appropriate the world around it for its own life."[1] The marvelous adaption of organisms to their environment is superb evidence of a Supreme Designer.

Comparative Anatomy

In the latter decades of the nineteeth century, evolutionary scientists labored mightily to establish evolution through the studies of comparative anatomy and comparative embryology.[2] Evolutionists still maintain that comparative anatomy is an important source of evidence for evolution.[3]

Comparative anatomy is the study of similarities between different kinds of animals. The term *homology* refers to the correspondence between the similar parts of different animals such as the arm of a man, the wing of a bird, the foreleg of a dog, and the pectoral fin of a fish.[4]

Homologous structures cannot prove evolution, as Dobzhansky acknowledges: "Homology does not prove evolution, in the sense that nobody has actually witnessed the gradual changes in the millions of consecutive generations which led from a common ancestor to a bird on the one hand and to man on the other."[5] Even so, evolutionists still maintain that the similarity in structure of different organisms is directly proportional to their closeness of kinship.

Many groups of animals are not closely related according to evolutionary theory, however, and yet have similar structures.[6] For example, whales and fish are very similar, but is this the result of close relationship or of a common creative design? No evolutionist claims it is the result of relationship, that is, one kind evolved from the other or both kinds had a recent common ancestor. Rather, the similarities are a result of common design. God designed both kinds to live in the water and endowed them with comparable features accordingly.

Evolutionists attempt to overcome this dilemma by developing a bracing theory to support their weakened primary theory of evolution. They call it "parallel evolution." In other words, two groups of organisms not closely related may have evolved similar structures because similar habitats would have afforded them essentially the same evolutionary opportunity.[7] If similarity in structures without possible relationships between organisms occurred only rarely in nature, then parallel evolution would be a plausible explanation. However, parallelisms occur in so many cases and among so many different groups of animals that it is now regarded as a universal phenomenon.

From the viewpoint of design, we could compare similar structures in different organisms to a set of encyclopedias. Every volume has a similar cover but each has different contents. We could say one volume evolved into the other on down through the entire set of encyclopedias, but in fact all the volumes had the same publisher. Certainly the arm of a man, the wing of a bird, the foreleg of a dog, and the pectoral fin of a fish are similar in structure, but is this the result of a common ancestor or a common creator?

Recapitulation Theory

The "recapitulation theory" was developed in 1866 by Ernst Haeckel. He called it the "biogenetic law," and expressed it in the words, "Ontogeny recapitulates phylogeny."[8] In other words, as individual organisms develop, their various embryonic stages supposedly correspond to ancestral adults.[9]

For many years evolutionary scientists universally accepted the recapitulation theory. As research continued in embryology, however, scientists discovered they could not apply the recapitulation theory exactly but had to make certain important exceptions and qualifications.[10] In fact, according to Hotton, "Haeckel's zeal carried him to speculations on evolutionary relationships among various creatures that were quite wrong and an embarrassment to Darwin."[11]

Today many scientists seriously question the recapitulation theory, and some high authorities have altogether denied it.[12] As evolutionist Newman comments:

> *The fascinating Biogenetic Law of Muller and Haeckel summed up in the phrase, "—ontogeny is a recapitulation of phylogeny," must not be too heavily leaned on as a support for any speculations as to the phyletic affinities of any species or group of species of organisms. "Embryology is an ancient manuscript with many of the sheets lost, others displaced, and with spurious passages interpolated by a later hand."*[13]

Some uninformed evolutionists still believe man goes through the history of his evolutionary past as he develops in embryonic form from a fertilized egg to a fully developed baby. They regard this as evidence that man could have evolved from a single-celled amoeba. What they fail to realize is that the fertilized egg of man possesses a genetic blueprint plus an outside energy source, which actually produces the baby, while the amoeba has neither.

This theory is comparable to placing bricks, lumber, steel, and other building materials in an open field and expecting them to evolve into a ten-story building over billions of years.

Although this analogy involves inanimate objects, evolutionary theory says the first organic materials evolved from inorganic materials. If the building materials remained in the open field for eternity, they still would not produce the building, for they are incapable of doing so independently. Actually time is working against this end for eventually the Second Law of Thermodynamics would begin to take its toll. The bricks would crack and fall apart, the lumber would rot, and the steel would rust.

This analogy is actually oversimplified. The body processes of man are in reality many times more complex than a ten-story building. If we increased the building to 100 stories and placed wall-to-wall computers in each room of every story, then the analogy would come closer to reality. We cannot give the dimension or number of rooms because no one knows just how complex man really is. Man is discovering new information about his complexity and functions every day. By the time this book is published, the rooms would already need to be expanded.

In reality, a human embryo develops into a human baby because that is what its genetic blueprint tells it to do, and an amoeba produces another amoeba because that is what its genetic make-up tells it to do.

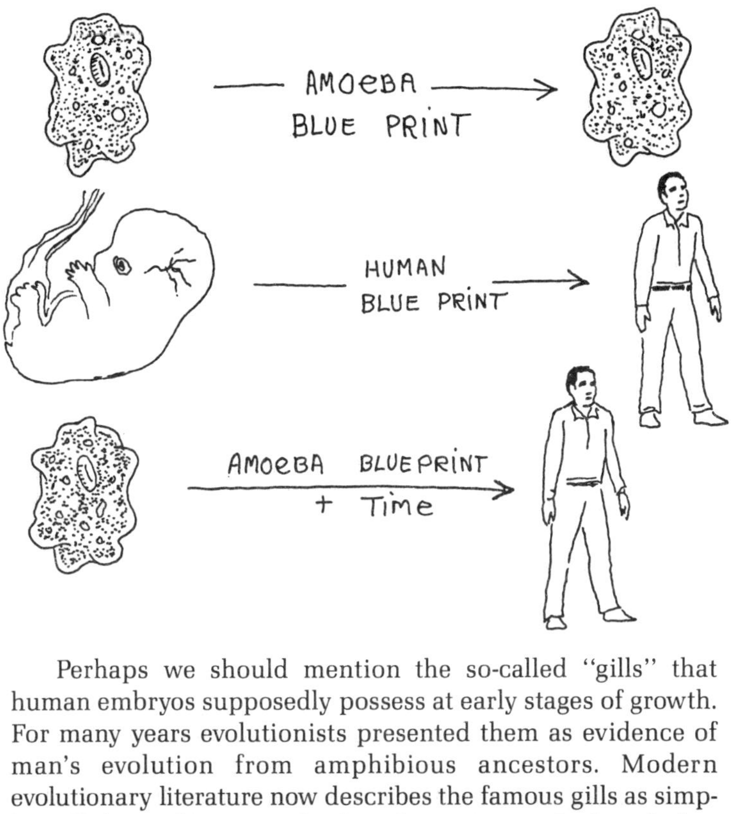

Perhaps we should mention the so-called "gills" that human embryos supposedly possess at early stages of growth. For many years evolutionists presented them as evidence of man's evolution from amphibious ancestors. Modern evolutionary literature now describes the famous gills as simply "gill slits." The reason for this change in terminology is that the human embryo does not have gills but rather has folds of tissue that develop into parts of the tongue, lower jaw, and neck.

These folds of tissue cannot by any stretch of the imagination be compared to gills. Again another "fact" has become fiction.

Classification

Biologists place animals and plants into categories of closest similarity. The major types of categories (from close to distant relationships) are species, genera, families, orders, phyla, and kingdoms. Evolutionists point to the similarities that allow living things to be classified as evidence for the theory of evolution, but this is actually one of the strongest indictments against the theory. If all life evolved from one source, there should be an intergrading of every plant and animal without any gaps whatsoever. However, this is not the case. There are numerous, clear-cut gaps which allow plants and animals to be classified.

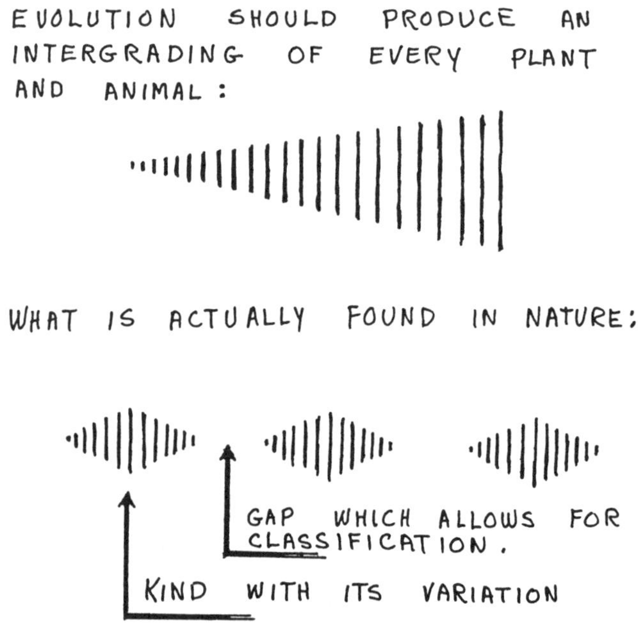

The evolutionary terms *primitive* and *advanced* are often used when classifying organisms. The determination of whether a feature is primitive or advanced is based solely upon the opinion of the researcher. If the researcher assumes evolution has proceeded in a certain direction, he will classify the features of the supposed "earlier stages" as more primitive than the features of the supposed "later stages."

This is another obvious example of circular reasoning. The evolutionist makes the assumption that evolution has followed a certain direction. He next uses the assumption as evidence that a feature is primitive or advanced. He completes the circle by using the primitive and advanced characteristics as evidence to support his original assumption. Often, the same characteristic that is termed "primitive" by one researcher will be termed "advanced" by another researcher.

Complex Organs

Even Charles Darwin noted the difficulty evolution has in explaining the origin of complex organs:

> *Is it possible that an animal having, for instance, the structure and habits of a bat, could have been formed by some other animal with widely different habits and structure? Can we believe that natural selection could produce, on the one hand an organ of trifling importance, such as the tail of a giraffe, which serves as a fly-flapper, and, on the other hand, an organ so wonderful as the eye?* [14]

If man is evolving, organs in his body should be evolving, but even Darwin knew that an incompletely formed organ is a hindrance to the body. For the process of natural selection to select a change in an organism and pass it on to succeeding organisms, the change must have immediate value to the organism. If not, then natural selection would carry it no farther in its struggle for existence. For this reason, the population of animals that possessed an incomplete organ would be least likely to survive.

This places evolutionists in a real dilemma, for they have no mechanism for the evolution of complex organs such as eyes, ears, and lungs; or mechanical parts of the body such as hands, feet, limbs, and wings. The very mechanism of natural selection proposed for evolution would eliminate any incomplete organ or limb from the body, for what value is an incomplete eye or an incomplete wing or any other organ that does not function?

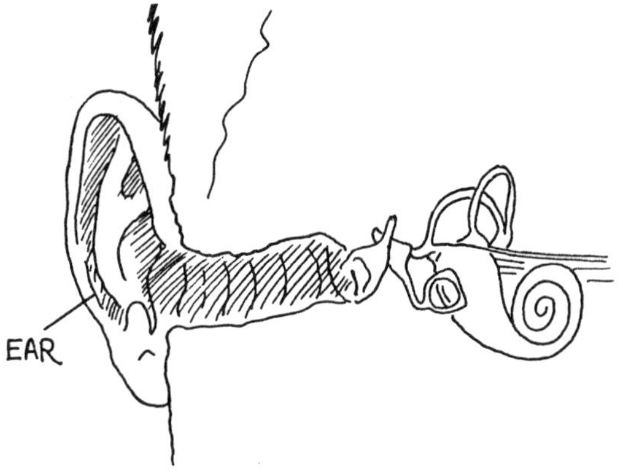

For the evolutionist, the existence of complex organs is an unsolvable mystery; from the viewpoint of design it is exactly what we would expect. God designed eyes to see, ears to hear, lungs to breathe, legs to walk, and wings to fly.

Darwin acknowledged:

> *To suppose that the eye with all its inimitable contrivances for adjusting the focus to different distances, for admitting different amounts of light, and for the correction of spherical and chromatic aberration, could have been formed by natural selection, seems, I freely confess, absurd in the highest degree.*[15]

Footnotes
Chapter Eight

[1]Richard C. Lewontin, "Adaptation," *Scientific American*, Vol. 239, No. 3 (Sept. 1978), p. 213.

[2]Dodson and Dodson, p. 430.

[3]*Ibid.*, p. 37.

[4]Cleveland P. Hickman and Cleveland P. Hickman, Jr., *Biology of Animals* (C. V. Mosby Company, 1972), p. 77.

[5]Dobzhansky, p. 227.

[6]Hickman and Hickman, p. 325.

[7]*Ibid.*

[8]Dodson and Dodson, p. 52.

[9]*Ibid.*

[10]Newman, p. 114.

[11]Hotton, p. 136.

[12]Newman, p. 114.

[13]*Ibid.*, p. 113.

[14]*Ibid.*, p. 590.

[15]*Ibid.*, p. 591.

Chapter 9
VARIATION

In our modern society, almost every product we purchase, from cars to can openers, is manufactured on assembly lines, which makes every item to be identical to the previous one. If there is a variation, the item is considered to have a flaw and the article is discarded or re-manufactured.

Not so in God's assembly line. Every individual plant or animal of a given species is unique in its own way. For example, no two trees of the same species are ever identical. In like manner, no two people are identical. Even so-called identical twins actually are not identical, but vary from one another. Every person, just as every plant or animal, has individual characteristics. As the psalmist David proclaimed in Psalm 139:14, "I will praise thee; for I am fearfully and wonderfully made: marvelous are thy works; and that my soul knoweth right well."

This book has been mass-produced by the thousands. If someone purchases another copy, it will have the same cover, the same number of pages, and the same number of words arranged in the same order. Once someone has read this book, it would be useless to buy another copy and hope to learn more from it than from the first copy. In God's creation, however, we see organisms millions of times more complex than this book, and each individual is different, unique, and beautiful. Scientists can study the same species of any plant or animal for years and learn something new from each specimen observed.

Today, there is much variation within the given "kinds" of organisms spoken of in Genesis 1 and 2. Much of this variation is the result of varying environmental factors. Before the Flood, there was not much variation in the environment; therefore, there probably were not as many varieties within a given "kind" as there are today.

Variation does not prove evolution, but to the contrary it strongly points toward creation. God recognized the need for organisms to adapt to new environments. Therefore he placed built-in mechanisms in the genetic make-up of organisms to enable them to populate many different environments.

There are definite limits of variation within a given "kind," with the extremes of variation being the genetic limits of the "kind." Each "kind" was created with a variety of genes that allow it to adapt to different environments, but once the limits of variation have been reached, there can be no further adaptation.

A good example of the limits of variation is found in food crops. For years, agriculturists have continuously worked to increase crop yield. At first, they made great strides, doubling and even tripling the yields. In the last few years, however, they have begun to reach the "limits of increased yield." What they should really say is they are approaching the outer limits of variation within each particular "kind" of food crop.

We should point out that one "kind" of organism is totally separate from another "kind" of organism. For example, there are many varieties of dogs within the "dog kind" and there are many varieties of cats within the "cat kind," but the two "kinds" are totally distinct from each other.

The Bible plainly states this principle in Genesis 1:25: "And God made the beast of the earth after his kind, and cattle after their kind, and every thing that creepeth upon the earth after his kind: and God saw that it was good." True science agrees with this totally. "Thus, in producing more of its kind, each organism passes on influences which will control the growth of the offspring and cause it to become like its parents. *Like produces like*—this is another characteristic of living things."[1] Yet, evolution totally contradicts this characteristic of all living matter.

Variation is a comparative term that can only be applied to members of the same "kind." An example is the finches of the Galapagos Islands, Darwin's most popular evidence for the theory of evolution. There is much variety in their bill size, coloration, and feeding and perching habits, yet they are all finches. This simply illustrates the great amount of variation possible within "kinds."

Some may argue that the various types of finches of the Galapagos Islands are classified as different species, yet Darwin himself said that a species in any group is whatever a competent specialist on that group says a species is.[2] Newman explains, "Those individuals which agree very closely in structure and function belong to the same species. There is no absolute test, other than the common judgment of men competent to decide."[3] Newman further states that there is no absolute definition for the word *species*.[4]

We can readily see that the concept of "species" is a very loose term. The various species of finches Darwin described could actually be one species or at least one "kind" in the meaning of Genesis 1 and 2.

Natural Selection and Mutations

Charles Darwin based his theory of evolution upon the mechanism of acquired characteristics acting through natural selection. For example, he believed that giraffes got their long necks by continually stretching their necks to reach food higher and higher in trees. Natural selection favored the long-necked giraffes because they had an advantage in competing with their shorter-necked predecessors. They passed on their long necks to their descendants. Eventually the long-necked giraffes totally displaced short-necked ones.

In the late 1880's, however, the concept of acquired characteristics was totally disproved. Scientists demonstrated that acquired characteristics were not inherited and therefore could not influence the next generation.

In the early 1900's, a Dutch botanist named Hugo DeVries presented the "mutation theory." Many scientists at that time incorporated the "mutation theory" with Darwin's theory of natural selection and presented this combination as the mechanism driving the evolutionary process. It was debated for a time and then decided to be inadequate.

The evolutionists were cautious to maintain that although they rejected the mechanism of evolution, it did not mean that they rejected the theory of evolution. Even though they had no mechanism for evolution, they maintained that evolution did occur. Newman asserts: "In contrast to the unity of opinion on the Law of Evolution is the wide diversity of opinion on the causes of evolution. In fact, the causes of evolution of life are as mysterious as the law of evolution is certain."[5]

The evolutionists of the early 1900's stressed that the mechanism for evolution would eventually be found. Consequently, for a lack of a better mechanism, evolutionists today have been forced to return to the idea that mutations directed by natural selection are the mechanism for evolution.

Evolutionists are aware that natural selection explains the survival of the fittest but fails to account for the arrival of the fittest.[6] At this point, evolutionists claim mutations play their part. They say mutations provide organisms with new advantageous characteristics enabling them to compete better with their environment.

A mutation is a sudden fundamental change in hereditary material, which makes an individual basically unlike its parents. Mutations usually occur when genes are damaged by radiation or by a malfunction during the combination of chromosomes from the two parents.

Unfortunately for them, evolutionists are hardpressed to provide an example of a good mutation in nature. It is even questionable as to whether there are good mutations in nature. Dobzhansky says mutations that make major improvements in the normal organization of the fruit fly—a highly studied insect in the field of genetics—are unknown.[7]

In fact, all scientists, whether creationists or evolutionists, are aware of the fact that a very high percentage of mutations are unquestionably deleterious. Evolutionist Merrell remarks, "Most of the mutations that occur are deleterious and recessive to the prevailing types of genes."[8] Merrell further states that 99.9% of all mutations are not beneficial to the organism.[9] Even experimentally-induced mutations have almost all (if not all) been detrimental. "Although the possibility of using radiation to induce useful mutations has been extensively investigated, especially in crop plants, almost all of the experimentally produced mutations have been deleterious."[10]

Even so-called "improvements" from mutations are at the most superficial. For example, a mutation that increases the yield in a crop is good for man, but what about for the plant? Will it improve its chances of survival in the wild? Usually mutations that are good as far as man is concerned produce some other weakness or malformity in the plant that would not allow it to compete in the wild, but rather would hasten the plant's ultimate death.

Another good example is the breeding of cows to yield more milk. Dobzhansky explains:

> *Primitive cattle breeds yield only two or three quarts of milk per cow per day; modern dairy breeds produce several times more (as much as 10,000 pounds of milk and 300 or more pounds of butterfat per cow per year). On the other hand, primitive cattle can live well with little care under harsh conditions, whereas their modern counterparts require special feeding and attention. Has, then, the selection practiced for centuries by cattle breeders improved or damaged the fitness of the domestic cattle?*[11]

It is doubtful, then, that there is such a thing as a good mutation. If not, then the evolutionists are left without a mechanism for evolution. Even if there are good mutations, they are so few, so far between, and so outnumbered by bad ones that they do not provide an adequate mechanism for evolution.

At this point we should discuss the peppered moths *Biston betularia* of England, considered by most evolutionists as the best example of a good mutation in nature. In fact, when H. B. D. Kettlewell first published his article on the peppered moths, he entitled it "Darwin's Missing Evidence."[12] This so-called evidence is often cited as one of the best examples of evolution in action.

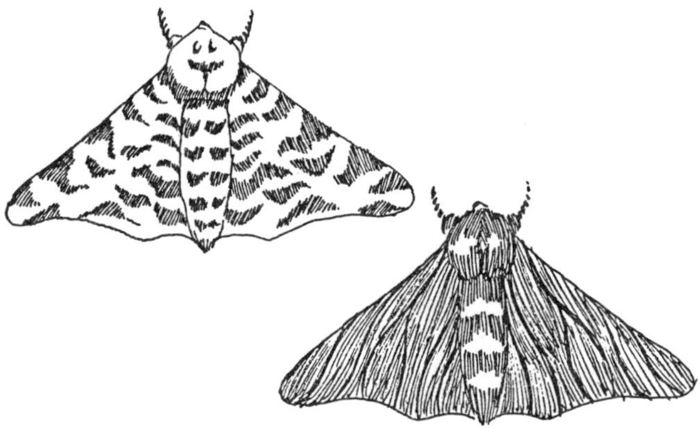

Before the industrial revolution there were white and black forms of the peppered moth in England. The whites greatly outnumbered the blacks, because the whites blended in with the light-colored tree trunks while the blacks were easily spotted by predators. As the tree trunks became black from factory smoke, the situation reversed, and blacks greatly outnumbered whites.

This example demonstrates natural selection or survival of the fittest, but not evolution. No new "kind" or even a new characteristic came into existence. Both blacks and whites had been there all along. All that changed was the ratio of white moths to black moths. This was merely change or variation within a "kind" rather than one "kind" evolving into another "kind." This change in ratio from white to black did not add to or take away from the species or "kind."

Ironically, this evidence is no longer so striking. In England today, many factories are cutting down on pollutants due to the environmental laws now in effect. For this reason the color ratio in the peppered moths is again reversing. The evidence for evolution is still missing.

Since mutations are so rare and since most, if not all of them, are degenerative, how could a beneficial mutation always appear on the scene at the right moment to push an organism a little higher up the evolutionary ladder? For a mutation to be beneficial, it has to occur when the environmental situation calls for it. If it were to happen at any other time, it would be of no benefit. More than likely it would then be harmful to the species and thus be eliminated from the gene pool by natural selection.

Furthermore, countless millions of mutations would be required to change one animal or plant "kind" into another "kind." Considering all this, what is the statistical probability that all the proper mutations would appear in time to advance the organism's fitness to cope with its changing environment? It certainly approaches zero.

At this point, an evolutionist would probably say the environment changed at such a slow rate over countless millions of years that there was ample time for a beneficial mutation to come along. If it takes great spans of time for a beneficial mutation to come along and it takes millions of mutations to produce a new kind, however, then even the 4.5 billion years given for earth's age is not enough time. It would not be enough time to produce even one new kind, much less enough time for an amoeba to evolve into a man.

Instead of natural selection being the mechanism by which species change, it is actually the mechanism by which species remain stable. Mutations are actually errors in the replication of DNA prior to its translation into protein.[13] These genetic malfunctions make the mutants less capable of competing with the normal members of their species: therefore, the process of natural selection eliminates these weaker organisms and in turn keeps the species pure.

Hybrids

Agriculturists and horticulturists spend millions of dollars per year in their endeavor to produce new varieties and hybrids of crops and ornamental plants. Evolutionists often cite this as an example of evolution today.

What these evolutionists fail to say is that these varieties are being produced by eliminating natural selection, a vital part of the mechanism of evolution. When new varieties are produced by hybridization, they are invariably weaker in one form or another than their wild parents. If these plants were placed under natural conditions, they would be incapable of competing with the wild types. This is why man-made hybrids, when placed back in their natural habitats, either revert back to their wild form through their progeny or die.

For example, let us imagine a poodle or a chihuahua in the wild trying to compete with a wolf or a coyote. Certainly the poodle and chihuahua would be eliminated in their struggle for existence.

Many plant and animal hybrids are sterile. Examples are mules and the domestic varieties of roses. These varieties and hybrids could not compete with their wild parents because after one generation they would be extinct.

In short, the existence of hybrids is evidence of variation within kinds and actually evidence against the possibility of evolution of one kind into another. Once again, evolutionary "fact" has become fiction.

Footnotes
Chapter Nine

[1]Truman Moon, James Otto, and Albert Towle, *Modern Biology* (Holt, Rinehart, and Winston, 1960), p. 14, emphasis in origina.

[2]Dodson and Dodson, p. 352.

[3]Newman, p. 97.

[4]*Ibid.*

[5]*Ibid.*, p. 7.

[6]*Ibid.*, p. 28.

[7]Dobzhansky, pp. 105-106.

[8]Merrell, p. 211.

[9]*Ibid.*, p. 350.

[10]Dodson and Dodson, p. 427.

[11]Dobzhansky, p. 122.

[12]H. B. D. Kettlewell, "Darwin's Missing Evidence," *Scientific American*, Vol. 200, No. 3 (March 1959).

[13]Francisco J. Ayala, *The Mechanisms of Evolution.*

Chapter 10
THE AGE OF THE EARTH

For evolution to be valid, there must have been a great expanse of time to produce the plants and animals now present in the biological community. Evolutionists believe that time, coupled with natural selection, has produced biological improvement.[1] Even Darwin stressed the importance of time as a factor in the theory of evolution.[2] As Dodson and Dodson point out, if the earth is young there has not been enough time for evolution to have occurred:

> *One of the outstanding accomplishments of James Hutton and Charles Lyell, the founders of modern geology, was to establish the great antiquity of the earth. If the earth were no older than the 6,000 years inferred from biblical genealogies, Darwin's studies on evolution would have been pointless, for the time necessary for*

evolution simply would not have been available.[3]

For this reason, evolutionists have gone to extremes in trying to prove that the earth is of great geological age.

Prior to Darwin's theory of evolution, most scientists were of the opinion that the earth was young in age, that is, probably no more than 6,000 years old. Due to the immense time span required for evolution, revolutionary scientists began to increase the possible age of the earth. In 1900 they proclaimed the earth to be 50 million years old. After more research, however, they decided 50 million years was not enough time to allow for the slow mechanism of evolution. Therefore, they have increased the presumed age of the earth to 4.6 billion years.[4] Even if the earth truly is millions of years old, this would not prove the theory of evolution: on the other hand, if the earth is only a few thousand years old, then the theory of evolution is completely disproven. With this in mind, let us examine some methods used in dating the earth.

Uranium-Lead Dating Method

The uranium-lead method of dating strata was developed in 1907. Since that time it has become the standard by which the accuracy of other methods of dating are judged.[5] The method is based upon the discovery that uranium 238, which is radioactive, slowly and constantly disintegrates to produce helium and lead 206.

The amount of time necessary for half the molecules of a specific amount of uranium to break down into lead and helium is called "the half-life" of uranium. The half-life of uranium has been calculated to be 4,510,000,000 years.[6] According to this, if a strata is found which contains uranium-bearing rock, then the ratio of uranium to lead can be determined, and thus the age of the rock formation can be calculated.

On the surface, the uranium-lead dating method would seem to yield very accurate results. Unfortunately, there are some very serious flaws in the method. First, it is impossible to calculate the amount of original lead deposited along with the uranium. In other words, it is impossible to know the original setting to the uranium-lead clock.

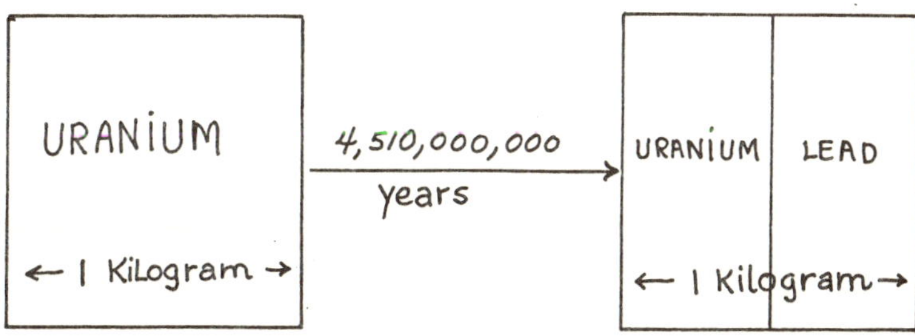

Second, it is possible for some uranium to leach out of the strata by acid waters. This gives readings of great age to strata that are in reality much younger.

Evolutionist G. A. Kerkut acknowledges that many dates obtained by the uranium-lead method are incorrect:

> *Some are very inaccurate due, it is believed, to the loss of radon by diffusion from the rocks in the series U^{238}/Pb^{206}. Another difficulty is due to the amount of non-radiogenic lead present in the material. Where this is high there is a corresponding high error in the estimation. This can lead to an error of 700 million years in the exceptional case of the Caribou Mine, Colorado, where the deposit contained as much as 97% lead.*[7]

Considering that the uranium-lead dating method is the standard by which the accuracy of other dating methods are judged, it is quite obvious that evolutionists are far from proving the antiquity of the earth.

Potassium-Argon Dating Method

Scientists once used the potassium-argon dating method with great hope of accuracy. However, they have recently discovered that incorrect age determinations could result (1) from the loss of argon from the strata, (2) by potassium-40 changing to calcium-40 rather than to argon, or (3) by contamination from the high argon content of the atmosphere. Even evolutionists now acknowledge the existence of very serious objections to the potassium-argon dating method.

Radiocarbon Dating Method

The radiocarbon dating method, commonly called the carbon-14 dating method, was developed for shorter range determinations of age, that is, up to a maximum range of 50,000 years.[8] In reality, it is accurate only to about 4,000 years.

Let us briefly discuss the principle by which the carbon-14 dating method works. Cosmic rays are constantly bombarding nitrogen-14 in the upper atmosphere, breaking down nitrogen-14 into radioactive carbon-14. The radioactive carbon-14 eventually unites with oxygen in the upper atmosphere to form radioactive carbon dioxide (CO_2). Plants and, indirectly, animals ingest the radioactive CO_2 along with the normal atmospheric CO_2. When the plant or animal dies it has a specific ratio of radioactive CO_2 to normal CO_2. The radioactive CO_2 slowly decays at a rate of one-half every 5760 ($\pm$30) years.[9] The longer the plant or animal has been dead, the less radioactive CO_2 will be in the ratio.

There is one major problem with the carbon-14 dating method. When Willard Libby first introduced the method in 1946, he assumed that radioactive CO_2 and normal CO_2 were in equilibrium in the atmosphere. To Libby this was a logical assumption. If the earth was billions of years old, the formation of radioactive CO_2 in the atmosphere and the rate of decay could have been in equilibrium for eons. It has now been calculated that the required amount of time needed to reach this equilibrium is merely 30,000 years.[10]

The works by Libby, Lingenfeter, Suess and others have clearly demonstrated that the formation of radioactive CO_2 is actually faster than its rate of decay.[11] In other words, the system is not yet in equilibrium. This yields an insurmountable problem for the theory of earth's great age because this disequilibrium between radioactive CO_2 and normal CO_2 in the atmosphere can only mean the earth's atmosphere is less than 30,000 years old. The very technique developed to defend the theory of evolution is now one of the many convincing arguments against it.

The Helium-4 Method of Dating the Atmosphere

Helium-4 enters the earth's atmosphere from three primary sources (1) as a by-product of the radioactive decay of uranium and thorium, (2) from cosmic helium, and (3) from cosmic ray-caused nuclear reactions in the earth's crust.[12]

Scientists have estimated that a 3×10^{11} grams of helium-4 enter the atmosphere per year. The atmosphere presently contains approximately 3.5×10^{15} grams of helium-4. Using this data, simple calculations will yield an approximate age of 12,000 years as the maximum age of the earth's atmosphere. This is, of course, a far cry from the four billion years evolutionists claim.

Many evolutionists contend there is so little helium-4 in the atmosphere because it somehow escapes into outer space. At this point, however, they must admit there is no plausible mechanism by which this escape could take place.

Dating by Means of Pressure in Oil Deposits

One of the strongest evidences for a young earth is the pressure found in oil deposits throughout the world. As Melvin Cook explains:

> *Such high pressures require sudden deep burial. Moreover, to retain them for such periods greater than 10,000 to 100,000 years is apparently impossible under the observed permeabilities [ability to permit leakage of fluids] of the [oil] reservoir and trap formation.*[13]

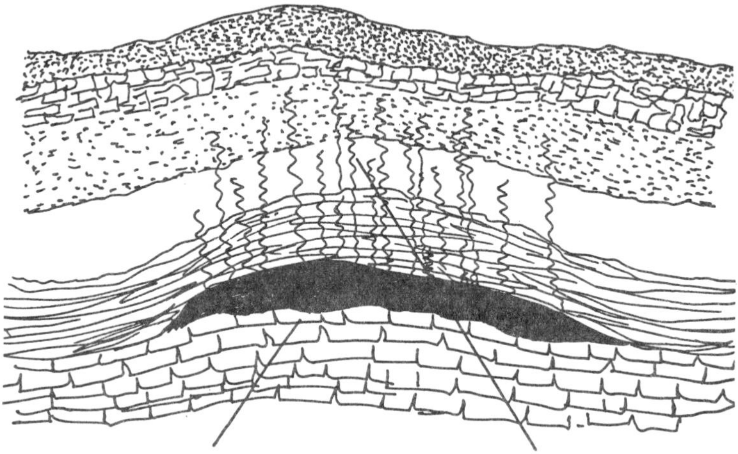

The very existence of pressure of any measurable amount, much less the extremely high pressures found in most deposits, is indisputable evidence that the oil deposits are a maximum of 100,000 years old and probably much younger.

Meteoritic Dust on the Earth's Surface

A large amount of meteoritic dust settles from the atmosphere to the earth's surface each year. Hans Pettersson of the Swedish Oceanographic Institute calculated this amount to be 14,300,000 tons per year.[14] If, as evolutionists say, the present is the key to the past, this means over 14 million tons of meteoritic dust have fallen to the earth each year for over 4.5 billion years: "If, through all that time, meteor dust had settled to the earth at the same rate it does today, then by now, if it were undisturbed, it could form a layer 54-feet thick over all the surface of the earth."[15]

Today, this 54-foot-thick layer of meteoritic dust is mysteriously missing from the surface of the earth. Asimov tries to explain its absence by suggesting that crustal mixing through the ages has caused a dispersion of the dust throughout the sedimentary layer. As creation scientists Morris and Whitcomb point out, however, this is impossible.

> *This means that the 54 feet thickness of meteoritic dust would have to have been dispersed through a crustal thickness of at least 312 x 54 feet, or more than three miles, to yield the present crustal nickel component percentage, even under the impossible assumption that there was no nickel in the crust to begin with.*[16]

Obviously, 54 feet of meteoritic dust could not have mixed in with the earth's crust for a thickness of three miles. Again, this evidence definitely presents a young age for the earth.[17]

Further Evidence

Further evidence supporting a young age for the earth includes the following:

1. Age of the earth based on rates of erosion and sedimentation.
2. Age of the earth based on volcanic activity.
3. Age of the oceans according to nitrate build-up.
4. Age of the sun.
5. Crustal accretion.
6. Dating of the moon's surface.
7. Dating of Saturn's rings.
8. Decay of the earth's magnetic field.
9. Disintegration of comets.
10. Juvenile water.
11. Meteorite radioactivity.
12. Population growth.
13. Salt content of the sea.
14. Tektites.
15. Tree ring dates.

The Literal Creation Week And Earth's Age

The Bible deals with three periods of time: (1) the past

(history), (2) the present (guide to salvation and daily living), and (3) the future (prophecies concerning events yet to come). We can have complete confidence in its validity in all three areas.

The Bible emphatically states in Genesis 1 and 2 that the earth was created in six days, with evening and morning comprising each day, and that God rested on the seventh day. Exodus 20:11 restates this truth: "For in six days the LORD made heaven and earth, the sea, and all that in them is, and rested the seventh day: wherefore the LORD blessed the sabbath day, and hallowed it." In Exodus 20:10, God commanded the people of Israel not to work on the Sabbath day. Unquestionably, the day spoken of in Exodus 20:10 was a literal 24-hour day, because when the people of Israel fulfilled this command they worked six literal days and rested one literal 24-hour day. The same Hebrew word for "day" (yom) appears in both Exodus 20:10 and 20:11, as well as throughout Genesis 1 and 2.

It is probable, then, that the six days of creation and the day of rest were literal 24-hour days. In addition, they were possibly a type or a foreshadowing of man's history upon the earth. The six days of work may typify 6,000 years of man's activity (work) upon the earth. The seventh day would then be a type of man's rest during the Millennium, the 1,000 years of Christ's reign on earth as recorded in Revelation 20:4-5.

The Old Testament Sabbath day was a type of the spiritual rest Spirit-filled believers enjoy today through the indwelling of God's Spirit. (See Isaiah 28:11-12; Acts 2:1-4, 38, 3:19; Hebrews 4:1-11.) Physically, God made man at the end of the sixth literal day of creation. Spiritually, God has been restoring fallen man for the last 6,000 years; man will not be spiritually complete until the return of Christ. (See I John 3:2.)

Ruin-Reconstruction Theory

The Ruin-Reconstruction Theory, also called the gap theory, has been around in one form or another for centuries.

Until the early 19th century it was not very popular because people generally interpreted the Bible literally and thus believed the earth to be young. In the early 1800's, however, scientists began to accept uniformitarian geology, which advocated a great age for the earth. In 1814, Thomas Chalmers of Scotland combined the theory of uniformitarianism with the ruin-reconstruction theory.[18] The footnotes of *Scofield Reference Bible* of 1917 have since made it enormously popular.[19]

According to the ruin-reconstruction theory, there is a gap between Genesis 1:1 and 1:2. God supposedly created the heavens, earth, and sea before the six days of creation recorded in Genesis 1. This original creation was destroyed in a worldwide catastrophe and then recreated basically as it is today. The fossils of plants, animals, and men in existence today are the remnants of the destruction of the original earth.

The theory seems quite plausible on the surface and could certainly account for the great age required by uniformitarian geology. As one looks deeper into the ruin-reconstruction theory, however, serious problems arise from both scriptural and scientific considerations. Here are four major problems with the theory:

1. According to the ruin-reconstruction theory, the earth was re-made during the six days of creation, leaving the fossilized remains of the original creation in or upon the earth. According to Exodus 20:11, "In six days the LORD made heaven and earth, the sea, and all that in them is." The Bible explicitly says God made everything in the earth in six days; therefore, the fossils could not have been produced prior to the creation week. Furthermore, if God destroyed all animals, vegetation, and (as some think) even men and then recreated them in six days, why did God not do the same in Noah's day? He could simply have destroyed the earth again and then could have recreated the plant and animal life. Instead, God commanded Noah to build an ark and to place within it representatives of all animals to repopulate the earth.

2. According to the ruin-reconstruction theory, all fossils in the geologic column represent animals that lived and died prior to the creation of Adam. Romans 5:12 states, however, "Wherefore, as by one man sin entered into the world, and death by sin; and so death passed upon all men, for that all have sinned." This verse of Scripture emphatically states that sin and death entered the world by Adam, thus making the death of animals impossible prior to the fall of Adam. Without death, then, no fossil could have formed prior to Adam.

3. The strongest scientific argument against the ruin-reconstruction theory is that it does not explain why fossils are found in any order of ascending complexity in the geologic column. Furthermore, the theory does not explain why so many plants and animals in the fossil record are similar or even identical to present-day plants and animals while others are radically different.

4. According to the ruin-reconstruction theory, the sun, moon, and stars were actually created prior to creation week rather than on the fourth day of creation as recorded in Genesis 1:14-19. A dense cloud blanket supposedly prevented their light from reaching the earth, and this blanket was not lifted until the fourth day. According to Genesis 1:6-8, however, the waters were not lifted into the atmosphere to form a vapor canopy until the second day of creation: "And God said, Let there be a firmament in the midst of the waters, and let it divide the waters from the waters. And God made the firmament, and divided the waters which were under the firmament from the waters which were above the firmament: and it was so. And God called the firmament Heaven. And the evening and the morning were the second day."

We see, then, from both Scripture and science that the ruin-reconstruction theory fails to harmonize the record of Scripture with a great age for the earth.

Day-Age Theory

The day-age theory originated in 1749 when the distinguished French scholar, Compte de Buffon, advanced the idea that the six days of creation may have been six long epochs of time.[20] The theory did not gain many followers until the 1800's. At this time theologians felt obligated to compromise the Word of God with the theory of uniformitarianism, which advocated a great age for the earth.

The day-age theory is based on the Hebrew word for day *(yom)*, which can mean a long period of time as well as the usual 24-hour period. Concerning this possibility, Arthur Custance sent letters to nine Hebrew scholars who were members of the faculties of nine leading universities; three in Canada, three in the United States, and three in England. Seven of the nine replied, and all stated that the word "day" used throughout the first chapter of Genesis means a literal 24-hour day.[21]

Alfred Rehwinkel further explains the meaning of *yom* in Genesis 1:

> *The first and literal meaning of the Hebrew word "yom" means a natural day, and unless the context forces us to assume another meaning, we must accept that literal first meaning. And more than that, when the Old Testament associated "yom", or day, with a definite numeral a solar of 24 hours is always meant; for example, Genesis 7:11; 8:14; 17:12 and Exodus 12:6. Only by doing violence to the first principle of hermeneutics is it possible to arrive at any other conclusion. Furthermore, as previously discussed, Exodus 20:9-11 leaves no doubt that the word "days" that is found in these verses is the Hebrew word "yamin" and occurs approximately 700 times in the Old Testament, and without exception always means literal days.*[22]

There are many problems the day-age theory cannot resolve. We will discuss three of the more interesting ones.

1. On the third day of creation, the earth brought forth

grasses, herbs, and fruit trees (Genesis 1:11-12). According to the theory of evolution, plant life appeared millions of years before animal life came on the scene. With each day representing an age, how did plants get pollenated if insects and animals did not appear until ages later?

2. On the fourth day of creation, God created the sun, moon and stars (Genesis 1:14-19). How did plants live if there was no sunlight on the third day? According to the day-age theory, there was an age when plants had no sunlight for photosynthesis. Some may explain, on the basis of Genesis 1:3, that light was available for photosynthesis before the creation of the sun and stars. If so, it still would not solve the problem, because there was an evening and a morning in each day. Each day-age would have contained a period of many years of light followed by a period of many years of darkness.

3. As with the ruin-reconstruction theory, the day-age theory requires death before Adam, which contradicts Romans 5:12.

Since the time of Darwin, people have sought to modify the biblical creation account, either by trying to read between the lines (as in the ruin-reconstruction theory) or by trying to stretch words to mean something else (as in the day-age theory). They are merely attempting to stay in accord with man-made theories. If they would not attempt to reconcile the Word of God with the theory of evolution, there would be no need to search Genesis 1 for evidence of a great age for the earth.

Theistic Evolution

Many Christians today hold that if evolution is eventually proven, then it only means God used the mechanism of evolution to create the world. This concept is called "theistic evolution."

People who believe in theistic evolution do not realize that in reality they are attacking the inspiration of the Bible, because the Bible clearly portrays man as a perfect being who fell into sin. For this reason God placed a curse upon him, and man has been degenerating ever since. By contrast, the theory of evolution portrays man as progressing upward.

From the scriptural standpoint, it is difficult for the theistic evolutionist to reconcile his theory to Genesis 1 and 2, because at the end of each creation day God said His creation was all good. Theistic evolutionists are forced to believe, however, that during the days of creation the processes of decay, struggle, suffering, natural selection, mutation, and predation were all in effect. Would God look down at violence, death, and destruction and say, "It is good"? This is totally contrary to the nature of God and to the teaching of Scripture in Romans 5:12. Furthermore, all the scientific evidence we have provided against evolution does away with theistic evolution as well as atheistic evolution.

If a person rejects the Old Testament account of creation and classifies Genesis 1 and 2 as mere fables, then he must also reject the New Testament, for the New Testament accepts as valid the creation, Adam and Eve, and the curse. If any portion of God's Word is in error, then we must reject the whole, including the message of salvation. God did not give us the Bible to reject and discard, but rather to accept and obey. We must always choose the latter course of action.

Footnotes
Chapter Ten

[1]Huxley, p. 54.

[2]Sol Tax, *Evolution After Darwin,* p. 11.

[3]Dodson and Dodson, p. 71.

[4]*Ibid.*, p. 72.

[5]*Ibid.*, p.73.

[6]*Ibid.*

[7]G. A. Kerkut, *Implications of Evolution* (Pergamon Press, 1960), pp. 138-139.

[8]Dodson and Dodson, p. 78.

[9]*Ibid.*

[10]Melvin Cook, *Radiological Dating and Some Pertinent Applications of Historical Interests—Do Radiological "Clocks" Need Repair?, in Scientific Studies in Special Creation*, ed. Walter Lammerts (Presbyterian and Reformed Publishing Co., 1971), p. 79.

[11]*Ibid.*, pp. 79-80.

[12]*Science and Research* (Creation-Science Research Center, 1973), p. 56.

[13]Melvin Cook, *Prehistory and Earth Models* (Max Parrish, 1966), p. 341.

[14]Isaac Asimov, "14 Million Tons of Dust Per Year," *Science Digest*, Vol. 45 (Jan 1959), p. 34.

[15]Asimov, p. 35.

[16]John Whitcomb, Jr. and Henry M. Morris, *The Genesis Flood* (The Presbyterian and Reformed Publishing Co., 1961), p. 380.

[17]Additional material on the scientific evidence for a young earth is available from Creation-Life Publishers, P. O. Box 15666, San Diego, California 92115.

[18]John Whitcomb, Jr., *The Early Earth* (Grand Rapids: Baker Book House, 1972), p. 115.

[19]Whitcomb, p. 116.

[20]Dunbar and Waage, p. 11.

[21]Raymond Surburg, "In the Beginning God Created," in *Darwin, Evolution, and Creation*, ed. Paul Zimmerman (Concordia Publishing House, 1959), p. 61.

[22]Henry Morris, *The Remarkable Birth of Planet Earth* (Institute for Creation Research, 1972), p. 87.

Bibliography

Anderson, J. Kerby and Harold Coffin. *Fossils in Focus.* Probe Ministries International, 1977.

Anikouchine, William and Richard Sternberg. *World Ocean, An Introduction to Oceanography.* Prentice-Hall, 1973.

Asimov, Isaac. "14 Million Tons of Dust Per Year," *Science Digest,* Vol. 45 (January 1959).

Ayala, Francisco, *The Mechanisms of Evolution.*

Barghoorn, Elso. *Evidence of Climatic Change in the Geologic Record of Plant Life.* Harvard University Press, 1953.

Burdick, Clifford. "Microflora of the Grand Canyon," *Annual,* Creation Research Society, Annual 3, No. 1, 1966.

Colbert, Edwin. *The Record of Climatic Changes as Revealed by Vertebrate Paleoecology.* Harvard University Press, 1953.

Cook, Melvin, "Formation and Dating of Oil and Coal Deposits," *Bible Science Newsletter,* Vol. VIII, (January 1970).

Cook, Melvin. "Radiological Dating and Some Pertinent Applications of Historical Interest—Do Radiological 'Clocks' Need Repair?" in *Scientific Studies in Special Creation,* ed. Walter Lammerts. Presbyterian and Reformed Publishing Company, 1971.

Cook, Melvin. *Prehistory and Earth Models.* Max Parrish, 1966.

Darwin, Charles. *The Origin of Species by Means of Natural Selection.* Crown Publishers, 1879.

Davidheiser, Bolton. *Evolution and Christian Faith.* Baker Book House Company, 1969.

Dickerson, Richard. "Chemical Evolution and the Origin of Life," *Scientific American,* Vol. 239, No. 3 (September 1978).

Dobzhansky, Theodosius. *Evolution, Genetics and Man.* John Wiley and Sons, 1955.

Dodson, Edward and Peter Dodson, *Evolution, Process and Product,* D. Van Nostrand Company, 1976.

Dodson, Edward and Peter Dodson. *Evolution, Process and Product,* 2nd ed. Litton Educational Publishing, 1976.

Dorf, Erling. "Climatic Changes of the Past and Present," *American Scientist,* Vol. 48, No. 3 (September 1960).

Dott, Robert, Jr. and Roger Batten. *Evolution of the Earth.* McGraw-Hill Book Company, 1971.

Dunbar, Carl and Karl Waage, *Historical Geology.* John Wiley and Sons, 1969.

Field, Richard, *Geology,* 4th ed. Barnes and Noble, 1934.

Fleming, Thomas. "Solving the Lost Continent Mystery," *Reader's Digest,* Vol. 113, No. 676 (August 1978).

Fuller, Harry and Zane Carothers. *The Plant World,* 4th ed. Holt, Rinehart and Winston.

Gallent, Roy. *Charles Darwin, The Making of A Scientist.* Doubleday and Company, 1972.

Gish, Duane. *Evolution, The Fossils Say No.* Creation Life Publishers, 1973.

Heinze, Thomas. *Creation vs. Evolution.* Baker Book House Company.

Hickman, Cleveland and Cleveland Hickman, Jr. *Biology of Animals.* C. V. Mosby Company, 1972.

Hotton, Nicholas III. *The Evidence of Evolution.* American Heritage Publishing Company, 1968.

Huxley, Julian. *Evolution in Action,* 1st ed. Harper and Brothers, 1953.

Huxley, Julian. *Evolution in Action,* rev. ed. The New American Library of World Literature, Signet Science Library Books, 1953.

Kerkut, G. A. *Implications of Evolution*. Pergamon Press, 1960.

Kettlewell, H. B. D. "Darwin's Missing Evidence," *Scientific American*, Vol. 200, No. 3 (March 1959).

Leakey, Richard. "Skull 1470," *National Geographic*, Vol. 143, No. 6 (June 1973).

Lewontin, Richard. "Adaptation," *Scientific American* Vol. 239, No. 3 (September 1978).

Merrell, David. *Evolution and Genetics*. Holt, Rinehart, and Winston, 1962.

Michener, James. "Life and Death of the Dinosaur," *Reader's Digest*, Vol. 105, No. 631 (November 1974).

Montagu, Ashley, *Man—His First Million Years*. World Publishing Company, 1957.

Moon, Truman, James Otto, and Albert Towle. *Modern Biology*. Holt, Rinehart, and Winston, 1960.

Moore, John. *Questions and Answers on Creation/Evolution*, Baker Book House Company.

Moore, L. R. "Some Sediments Closely Associated with Coal Seams," in *Coal and Coal-Bearing Strata*, eds. Duncan Murchison and Stanley Westoll. American Elsevier Publishing, 1968.

Morris, Henry. *The Remarkable Birth of Planet Earth*. Institute for Creation Research, 1972.

Nelson, Byron. *The Deluge Story in Stone*. Bethany Fellowship Publishers, 1968.

Newman, Horatio. *Evolution, Genetics, and Eugenics*, 3rd. ed. The University of Chicago Press, 1932.

Opik, Ernst. "Climate and the Changing Sun," *Scientific American* (June 1958).

Ostrum, John. "A New Look at Dinosaurs," *National Geographic*, Vol. 154, No. 2 (August 1978).

Patten, Donald. *The Biblical Flood and the Ice Epoch*. Pacific Meridian Publishing Company, 1966.

Pelczar, Michael and R. D. Reid. *Microbiology*. McGraw-Hill Book Company, 1958.

Prout, Timothy. "Has Evolution Stopped?" *Science Digest*, Vol. 48, No. 1 (July 1960).

Rastall, R. H. "Geology," *Encyclopedia Brittanica*, Vol. 10, 1956.

Science and Creation. Creation Science Research Center, 1973.

Science and Research. Creation Science Research Center, 1973.

Segraves, Kelly. *The Great Dinosaur Mistake*. Kelly L. Segraves Beta Books, 1975.

Shapley, Harlow. *Climate Change*. Harvard University Press, 1953.

Simpson, George. *Horses*. Oxford University Press, 1951.

Stoler, Peter. "Puzzling Out Man's Ascent," *Time Magazine*, Vol. 110, No. 19 (November 1977).

Surburg, Raymond. "In The Beginning God Created" in *Darwin, Evolution, and Creation*, ed. Paul Zimmerman. Concordia Publishing House, 1959.

Tax, Sol. *Evolution After Darwin*. The University of Chicago Centennial, n.d.

Tax, Sol and Charles Callender. *Issues in Evolution*. The University of Chicago Press, 1960.

Wabhofsky, Fred. "Noah, The Flood, The Facts," *Reader's Digest*, Vol. III, No. 665 (September 1977).

Waechter, John. *Prehistoric Man*. Octopus Books, 1977.

Washburn, Sherwood. "The Evolution of Man," *Scientific American*, Vol. 239, No. 3 (September 1978).

Whitcomb, John, Jr. *The Early Earth*. Baker Book House Company, 1972.

Whitcomb, John, Jr. and Henry Morris, *The Genesis Flood.* The Presbyterian and Reformed Publishing Company, 1961.

Wilder, H. H. *The Pedigree of the Human Race.* Henry Holt and Company, 1926.

The World Book Encyclopedia, Vol. 6. Field Enterprises Educational Corporation, 1960.